Advances in Intelligent Systems and Computing

Volume 405

Series editor

About this Series

The series "Advances in Intelligent Systems and Computing" contains publications on theory, applications, and design methods of Intelligent Systems and Intelligent Computing. Virtually all disciplines such as engineering, natural sciences, computer and information science, ICT, economics, business, e-commerce, environment, healthcare, life science are covered. The list of topics spans all the areas of modern intelligent systems and computing.

The publications within "Advances in Intelligent Systems and Computing" are primarily textbooks and proceedings of important conferences, symposia and congresses. They cover significant recent developments in the field, both of a foundational and applicable character. An important characteristic feature of the series is the short publication time and world-wide distribution. This permits a rapid and broad dissemination of research results.

More information about this series at http://www.springer.com/series/11156

Jezreel Mejia · Mirna Muñoz
Álvaro Rocha · Jose Calvo-Manzano
Editors

Trends and Applications in Software Engineering

Proceedings of the 4th International Conference on Software Process Improvement CIMPS'2015

Editors
Jezreel Mejia
Mathematics Research Center
Research Unit Zacatecas
Zacatecas
Mexico

Mirna Muñoz
Mathematics Research Center
Research Unit Zacatecas
Zacatecas
Mexico

Álvaro Rocha
DEI/FCT
University of Coimbra
Coimbra
Portugal

Jose Calvo-Manzano
Computer Languages and Systems
University of Polytechnical of Madrid
Madrid
Spain

ISSN 2194-5357 ISSN 2194-5365 (electronic)
Advances in Intelligent Systems and Computing
ISBN 978-3-319-26283-3 ISBN 978-3-319-26285-7 (eBook)
DOI 10.1007/978-3-319-26285-7

Library of Congress Control Number: 2015953796

Springer Cham Heidelberg New York Dordrecht London

Printed on acid-free paper

Springer International Publishing AG Switzerland is part of Springer Science+Business Media (www.springer.com)

Preface

This book contains a selection of papers accepted for presentation and discussion at The 2015 International Conference on Software Process Improvement (CIMPS'15). This Conference had the support of the CIMAT A.C. (Mathematics Research Center/Centro de Investigación en Matemáticas), FIMAZ (Faculty of Informatic/Facultad de Informática Mazatlán, México), AISTI (Iberian Association for Information Systems and Technologies/Associação Ibérica de Sistemas e Tecnologias de Informação), ROOKBIZ Solutions, México, ReCIBE (Revista electrónica de Computación, Informática, Biomédica y Electrónica) and ROPRIN (Optimization Network of Industrial Processes/Red de Optimización de Procesos Industriales). It took place at Faculty of Informatic Mazatlán of the Autonomous University of Sinaloa, México, from 28 to 30 October 2015.

The International Conference on Software Process Improvement (CIMPS) is a global forum for researchers and practitioners that present and discuss the most recent innovations, trends, results, experiences, and concerns in the several perspectives of Software Engineering with a clear relationship but not limited to software processes, Security in Information and Communication Technology, and Big Data Field. One of its main aims is to strengthen the drive toward a holistic symbiosis among academy, society, industry, government, and the business community promoting the creation of networks by disseminating the results of recent research in order to align their needs. CIMPS'15 built on the successes of CIMPS'12, CIMPS'13, and CIMPS'14, which took place in Zacatecas, Zac, México.

The Program Committee of CIMPS'15 was composed of a multidisciplinary group of experts and those who are intimately concerned with Software Engineering and Information Systems and Technologies. They have had the responsibility for evaluating, in a 'blind review' process, the papers received for each of the main themes proposed for the Conference: Organizational Models, Standards and Methodologies, Knowledge Management, Software Systems, Applications and Tools, Information and Communication Technologies and Processes in non-software domains (Mining, automotive, aerospace, business, health care, manufacturing, etc.) with a demonstrated relationship to software process challenges.

CIMPS'15 received contributions from several countries around the world. The papers accepted for presentation and discussion at the Conference are published by Springer (this book) and by another e-book. Extended versions of the best selected papers will be published in relevant journals, including SCI/SSCI and Scopus indexed journals.

We acknowledge all those who contributed to the staging of CIMPS'15 (authors, committees, and sponsors); their involvement and support is very much appreciated.

Mazatlán
Sinaloa
México
October 2015

Jezreel Mejia
Mirna Muñoz
Álvaro Rocha
Jose Calvo-Manzano

Organization

Conference

General Chairs

Jezreel Mejia, Mathematics Research Center, Research Unit Zacatecas, MX
Mirna Muñoz, Mathematics Research Center, Research Unit Zacatecas, MX

The two general chairs are researchers in Computer Science at the Research Center in Mathematics, Zacatecas, México. Their research field is Software Engineering, which focuses on process improvement, multi-model environment, project management, acquisition and outsourcing process, solicitation and supplier agreement development, agile methodologies, metrics, validation and verification, and information technology security. They have published several technical papers on acquisition process improvement, project management, TSPi, CMMI, and multi-model environment. They have been members of the team that has translated CMMI-DEV v1.2 and v1.3 to Spanish.

General Support

CIMPS General Support represents centers, organizations, or networks. These members collaborate with different European, Latin America, and North America Organizations. The following people have been members of the CIMPS conference since its foundation for the last 4 years.

Cuauhtemoc Lemus Olalde, Cimat Unit Zacatecas, MX
Angel Jordan, International Honorary, Software Engineering Institute, US
Laura A. Ruelas Gutierrez, Government Zacatecas, MX
Gonzalo Cuevas Agustin, Politechnical University of Madrid, SP
Tomas San Feliu Gilabert, Politechnical University of Madrid, SP

Local Committee

CIMPS established a local committee of selected experts from Faculty Informatic of Autonomous University of Sinaloa, México. The list below comprises the Local Committee members.
Manuel Iván Tostado Ramírez, Local Chair, MX
Alma Yadira, Quiñonez Carrillo, Local Co-chair, MX
Rosa Leticia Ibarra Martínez, Public Relations, MX
Delma Lidia Mendoza Tirado, Public Relations, MX
Carmen Mireya Sánchez Arellano, Finance, MX
Bertha Elena Félix Colado, Finance, MX
Héctor Luis López López, Logistics, MX
Laura Patricia Sedano Barraza, Logistics, MX
Miguel Ángel Astorga Sánchez, Staff, MX
Diana Patricia Camargo Saracho, Staff, MX
Oscar Manuel Peña Bañuelos, Conferences, MX
Ana María Delgado Burgueño, Conferences, MX
Rogelio Alfonso Noris Covarrubias, Webmaster, MX
Rogelio Estrada Lizárraga, Scientific Local Committee, MX
Ana Paulina Alfaro Rodríguez, Scientific Local Committee, MX
Lucio Guadalupe Quirino Rodríguez, Scientific Local Committee, MX
Alán Josué Barraza Osuna, Scientific Local Committee, MX
Rafael Mendoza Zatarain, Scientific Local Committee, MX
José Nicolás Zaragoza González, Scientific Local Committee, MX
Sandra Olivia Qui Orozco, Scientific Local Committee, MX

Scientific Program Committee

CIMPS established an international committee of selected well-known experts in Software Engineering who are willing to be mentioned in the program and to review a set of papers each year. The list below comprises the Scientific Program Committee members.
Adriana Peña Perez-Negron, University of Guadalajara, MX
Alejandro Rodríguez Gonzalez, Politechnical University of Madrid, SP
Alma Maria Gómez Rodriguéz, University of Vigo, SP
Antoni Lluis Mesquida Calafat, University of Islas Baleares, SP
Antonia Mas Pichaco, University of Islas Baleares, SP
Antonio Pereira, Polytechnic Institute of leiria, PT
Antonio de Amescua Seco, University Carlos III of Madrid, SP
Arturo Méndez Penín, University of Vigo, SP
Carla Pacheco, Technological University of Mixteca, Oaxaca, MX
Carlos Lara Álvarez, CIMAT Unit Zacatecas, MX

Diego Martín de Andrés, University Carlos III of Madrid, SP
Edrisi Muñoz Mata, CIMAT Unit Zacatecas, MX
Edwin León Cardenal, CIMAT Unit Zacatecas, MX
Elisabet Cápon, Swiss Federal Institute of Technology, Zürich (ETHZ), CH
Enrique Barreiro Alonso, University of Vigo, SP
Fernando Moreina, University of Portucalense, PT
Francisco Ortín Soler, University of Oviedo, SP
Giner Alor Hernandez, Technological University of Orizaba, MX
Gloria P. Gasca Hurtado, University of Medellin, CO
Gonzalo Luzardo, Higher Polytechnic School of Litoral, EC
Gustavo Illescas, National University of Central Buenos Aires Province, AR
Henrique Mamede, Open University of Lisbon, PT
Hugo Arnoldo Mitre, CIMAT Unit Zacatecas, SP
Ivan A. Garcia Pacheco, Technological University of Mixteca, Oaxaca, MX
Jaime A. Guzmán Luna, National University of Colombia, CO
Javier García Guzman, University Carlos III of Madrid, SP
Joao Barroso, University of Trás-os-Montes and Alto Douro, PT
Jörg Thomaschewski, University of Vigo -Hochschule Emden-Leer, SP
Jose A. Mora Soto, CIMAT Unit Zacatecas, MX
Jose Antonio Cerrada Somolinos, National Distance Education University, SP
Josue N. Garcia Matias, Nova Universitas SUNEO, MX
Jose Baltasar García Perez-Schofield, University of Vigo, SP
Juan Carlos Gonzáles Moreno, University of Vigo, SP
Juan Manuel Cueva Lovelle, University of Vigo, SP
Leandro Rodríguez Linares, University of Vigo, SP
Luis Casillas, University of Guadalajara CUCEI, MX
Luis Fernández Sanz, University of Alcalá, MX
Luis Borges Goveia, University Fernando Pessoa, PT
Luis J. Dominguez Pérez, CIMAT Unit Zacatecas, MX
Magdalena Arcilla Cobián, National Distance Education University, SP
Manuel Pérez Cota, University of Vigo, SP
María José Lado Touriño, University of Vigo, SP
María Y. Hernández Pérez, Electrical Research Institute, MX
Marion Lepmets, Regulated Software Research Centre, Dundalk Institute of Technology, Ireland
Omar S. Gómez, Higher Polytechnic School of Chimborazo, EC
Paul Clarke Dundalk Institute of Technology, IE
Perla Velasco-Elizondo, Autonomous University of Zacatecas, MX
Rafael Valencia García, University of Murcia, SP
Ramiro Goncalves, University of Trás-os-Montes and Alto Douro, PT
Raúl Aguilar Vera, Autonomous University of Yucatán, MX
Ricardo Colomo Palacios, Østfold University College, NO
Rory O´Connor, Dublin City University, IE
Santiago Matalonga, University ORT, UY
Sodel Vázquez Reyes, Autonomous University of Zacatecas, MX

Sulema Torres Ramos, University of Guadalajara CUCEI, MX
Ulises Juárez Martínez, Technological University of Orizaba, MX
Ulrik Brandes, University of Konstanz, GL
Valentine Casey, Dundalk Institute of Technology, IE
Vianca Vega, Catholic University of North Chile, CL
Victor Saquicela, University of Cuenca, EC
Vítor Filipe, University of Trás-os-Montes and Alto Douro, PT
Yilmaz Murat, Çankaya University, TR

Contents

Part I
Organizational Models, Standards and Methodologies

Project Management in Small-Sized Software Enterprises: A Metamodeling-Based Approach

I. Garcia, C. Pacheco, M. Arcilla and N. Sanchez

Abstract Software development involves a unique effort that comprises managing many activities, resources, skills, and people to build a quality product. Thus, this effort is frequently seen from two different perspectives: the software development perspective and the project management perspective. Nowadays, any software enterprise that aims to develop high quality products should perform an adequate combination of both perspectives. However, such integration is generally not well addressed by the small-sized software enterprises due to the lack of knowledge, resources and time. In this sense, this paper introduces a metamodel to define a "lite" version of the project management process and to manage the generated knowledge during the software development.

Keywords Software process improvement · Project management · Metamodel · Small-sized software enterprises · Small teams

I. Garcia (✉) · C. Pacheco · N. Sanchez
Division de Estudios de Posgrado, Universidad Tecnologica de la Mixteca, Carretera a Acatlima, 69000 Huajuapan de León, OAX, Mexico
e-mail: ivan@mixteco.utm.mx

C. Pacheco
e-mail: leninca@mixteco.utm.mx

N. Sanchez
e-mail: nsanchez@mixteco.utm.mx

M. Arcilla
Escuela Técnica Superior de Ingenieros Informáticos, Universidad Nacional de Educación a Distancia, Ciudad Universitaria, 28040 Madrid, Spain
e-mail: marcilla@issi.uned.es

J. Mejia et al. (eds.), *Trends and Applications in Software Engineering*, Advances in Intelligent Systems and Computing 405,
DOI 10.1007/978-3-319-26285-7_1

1 Introduction

The small-sized software enterprises—companies with less than 50 employees that have been independently financed and organized—make a significant contribution to the economy of any country in terms of employment, innovation and growth. According to [1], project management can play a significant role in facilitating this contribution, but this kind of enterprises require less bureaucratic ways of projects management than those used by larger organizations. More importantly, in the US, Brazil, Canada, China, Mexico, India, Finland, Ireland, Hungary, and many other countries, small companies represent up to 85 % of all software organizations [2]. However, to persevere and grow, the small-sized software enterprises need efficient and effective software engineering solutions. Unlike large companies, the small ones don't have enough staff to develop functional specialties that would enable them to perform complex and secondary tasks to improve the quality of their software products. In this sense, it is true that some project management procedures/methods for medium-sized enterprises have been developed and these also have paid attention in the human aspects of Software Engineering (e.g., teamworking). Even so, they are also too bureaucratic for the smallest enterprises. Besides, in the context of these enterprises, it is common that software projects are developed by *small teams*[1] that are frequently integrated by poorly-skilled practitioners that have to be able to do all the work. Moreover, small-sized software enterprises require a "micro-lite" version of project management to support this way of work, with a preference for laissez-faire styles of management [4]. This is difficult, but not impossible.

For example, traditional methodologies usually provide guidance on what steps to follow in order to manage a project for obtaining a desired product. However, an extra effort is required when trying to match a given methodology with the small enterprise's enacted processes. But, what would happen if a small team receives a simplification of the practice? In this sense, metamodels have been proposed as a mean for proper understanding of the methodologies/processes through modeling. In essence, using metamodels means modeling a methodology as if it were any other system, applying the same modeling ideas and procedures that are usually applied to business applications or other software-intensive systems [5]. According to Callegari and Bastos [6], a metamodel can provide a conceptual model that leads to a formalization of concepts, so, formal definitions of the process can be used to generate tools for supporting the software development process. Thus, this paper aims to present a metamodel that represents, in this context, a simplified version of the basic project management activities. The rest of this paper is organized as follows: Sect. 2 outlines related work in the use of metamodels in Software Engineering. In Sect. 3 our metamodel is presented in order to provide an easy

[1]A small team is composed by a maximum of 10 people who usually work with a very limited infrastructure and resources. Nevertheless, these teams can be more productive than the large ones, because they are rapidly integrated and they usually work with less communication problems [3].

description for project management. Furthermore, in order to illustrate the practicability of this metamodel a brief description of an add-in for Microsoft Project® 2007 is also provided. Finally, Sect. 4 draws the conclusions and suggestions for future research.

2 Related Work

In the context of Software Engineering research, a recent analysis presents the relevance of study how people manage the software development process [7]. In this sense, 20 years ago Feiler and Humphrey had stated that any form of management "should be tailored to the needs of the project" [8], a concept that has been widely studied. Nevertheless, these needs are influenced by the situational context wherein the project must operate and therefore, the most suitable software development process should be contingent on that context [9]. Thus, project and process managers must evaluate a wide range of contextual factors before deciding on the most appropriate process to adopt for any given project [10, 11]. Based on these arguments some researchers have been creating descriptive models for documenting and, thereby, better understanding implementations of project management activities. Such models enable the better study and evaluation of project management and can thereby bring a finer granularity for the understanding of the practice [12]. Thus, the benefit to small-sized software enterprises from using these models is to get a better understanding of the project management activities by obtaining a "simplified" version of the actual practice. In this sense, a metamodel is also a model and it is generally defined as a "model of models" or, equally, "a model of a set of models" [5]. Therefore, metamodels can be used in order to increase the general understanding of the enterprise systems and processes and, specifically, to perform various kinds of analysis.

In this context, the use of metamodels has been increasingly investigated by software engineering researchers to simplify the software process. However, in spite of diverse areas of Software Engineering have been explored under the metamodel approach (e.g., requirements engineering [13–15], software process assessment [16, 17] and measurement [18, 19], software process improvement [20–22]), the project management has been poorly analyzed. For example, in [23] the PROMONT ontology for project management was introduced to build a common understanding of project related terms and methods, and thus, facilitating the management of projects conducted in dynamic virtual environments. Moreover, a workflow for teaching and practicing the agile project management methods by using LEGO® bricks is presented in [24]. This workflow is a result of the metamodel of the methods—agile project management and bricks building, and suggests new more creative ways of playing that are in fact learned from agile software development. Similarly, research by Callegari and Bastos [6] presented a metamodel for software project management based on PMBOK® Guide and its integration with RUP (Rational Unified Process). The purpose of this research was to provide a complete understanding of the core concepts of both models, providing a mapping between them, and proposing an integrated model that can guide practitioners towards a better management and, consequently, better products. Finally, Mas and Mesquida [25] recommend a set of best practices to facilitate the implementation of project management processes in small-sized software enterprises. This research consists on analyzing the relationships between different standards containing project management processes. As a conclusion, the authors proposed to use the ISO/IEC 29110-5-1-2 Project Management process and complement it by the knowledge of the selected areas of the PMBOK® Guide.

In conclusion, the research on metamodels to reduce the complexity of project management has been quietly developed and mainly focused on small enterprises. Continuing with this approach, we provide a metamodel for supporting the definition of a "lite" version of the project management process by promoting the knowledge management.

3 Defining a Metamodel for Project Management

According to Henderson-Sellers [5], metamodeling in current Software Engineering follows one of two possible architectures (see Fig. 1). The OMG (*Object Management Group*) architecture that is based on strict metamodeling [26] wherein the only relationship between levels is called "instance of" (left side of Fig. 1). In OMG standards, for example, an M0 object is said to be an instance of a class in level M1; a class in level M1 is said to be an instance of a metaclass in level M2 and so on. On the other hand, the right side of Fig. 1 shows an alternative multi-level architecture that introduces the *powertype* pattern as used in ISO/IEC 24744 [27]. Thus, an object facet provides attributes for Method Domain entities while the class facet provides specification of attributes that are given a value in the Endeavour

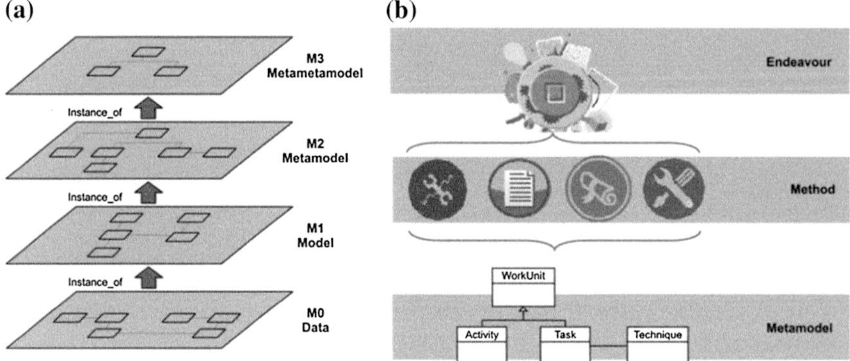

Fig. 1 Architectures for metamodeling in current software engineering [5]

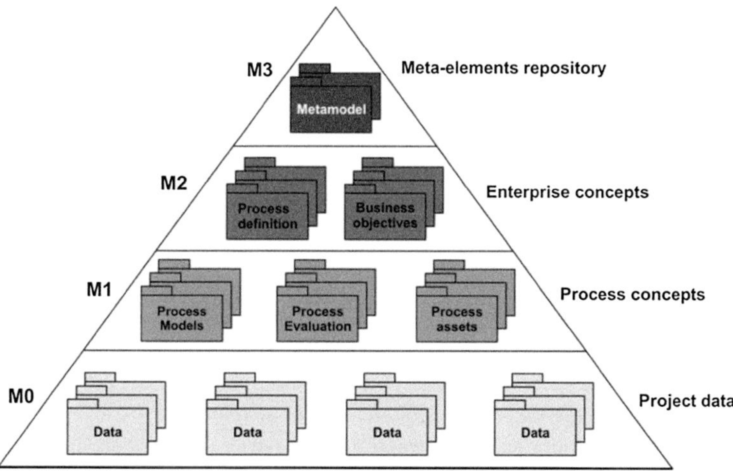

Fig. 2 The conceptual architecture for the metamodel

Domain. Furthermore, the use of *powertypes* permits both instance-of and generalization relationships between levels. Thus, in the context of defining methodologies for software development, this pattern combines the main advantages of other metamodeling approaches and enables the integration of documental aspects into the methodology [28].

In this sense, taking into account the ISO/IEC 24744 recommendations, a conceptual architecture has been proposed to build our metamodel (see Fig. 2). This representation aims to integrate all the process-related elements through the four layers and leads the formalization of concepts in the small enterprise context.

Thus, the purpose of this conceptual architecture is providing a mechanism to improve the understanding and use of a process through the four layers of abstraction:

- Level M0: The data generated by the projects represent the "real word" in the architecture. Thus, all historical data obtained by developing successful or failed projects are useful to learn how to manage them and, in a long-term, to predict the process performance.
- Level M1: An adjustment of complexity of practices recommended by process reference models is needed to provide to small enterprises a simplification of the practice. In this sense, our metamodel provides a set of scripts, templates and guidelines (assets) to support the project manager's labor. Additionally, the process evaluation is a crucial activity for this level, because it is not possible to define (adapt) a new process without determining the current state of practice (weaknesses and strengths).
- Level M2: The definition of a new process is related to the necessity to adequate a simplified description in a specific context. The pattern concept explored by González-Pérez and Henderson-Sellers [28] is introduced in M2 for enabling

project managers to take, from the concepts defined in M1, those useful to develop a specific project. This level uses a process asset library to enable project managers the definition of an enacted process.

- Level M3: The repository of all previous meta-elements uses instances at different levels, as an analogy to the SPEM metamodel [29], to create different versions of a process from other processes that may be previously defined.

Thus, following this conceptual model, the metamodel shown in Fig. 3 is presented with the intention to facilitate the definition of project management in small-sized software enterprises. It is important to mention that this metamodel summarizes the essential elements of project management that a small enterprise can use to begin to manage their projects. Furthermore, the main idea of this research is that with this metamodel a small enterprise can begin to mature at project-level and, in the future, mature at process-level.

To explain the metamodel, our description initiates from "basic" concepts and then presents their relationships to define a simplified managing schema. Due to space reasons, the details (such as all metamodel's attributes) were not included, but only the elements necessary for the explanation of this paper. Through the metamodel, the ***enterprise*** develops a ***project*** by integrating teams that perform activities by following, or trying to follow at least, a ***methodology***. Regularly a methodology states "what to do" in software development, but not "how", neither "who does it". Thus these methodologies are composed of a collection of ***phases***. Each phase is

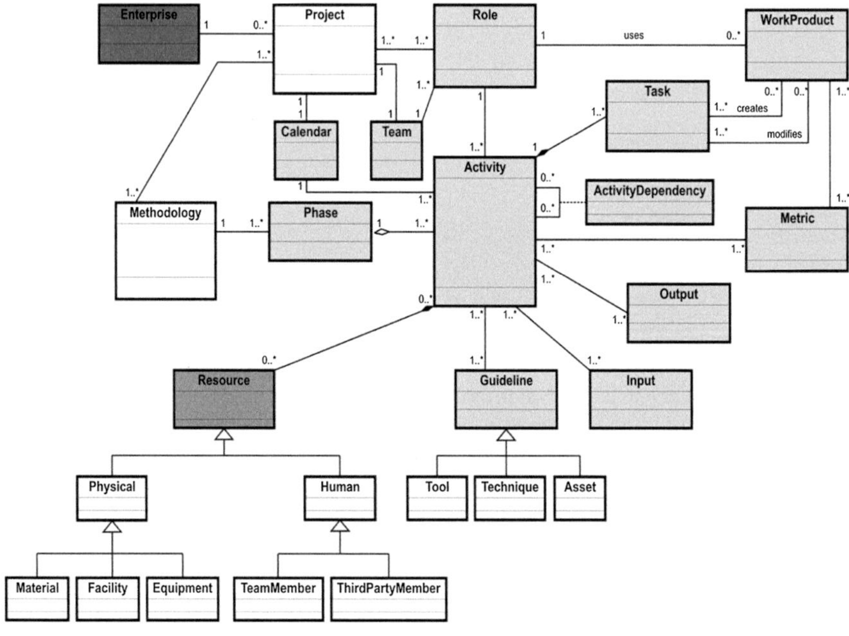

Fig. 3 The metamodeling-based approach

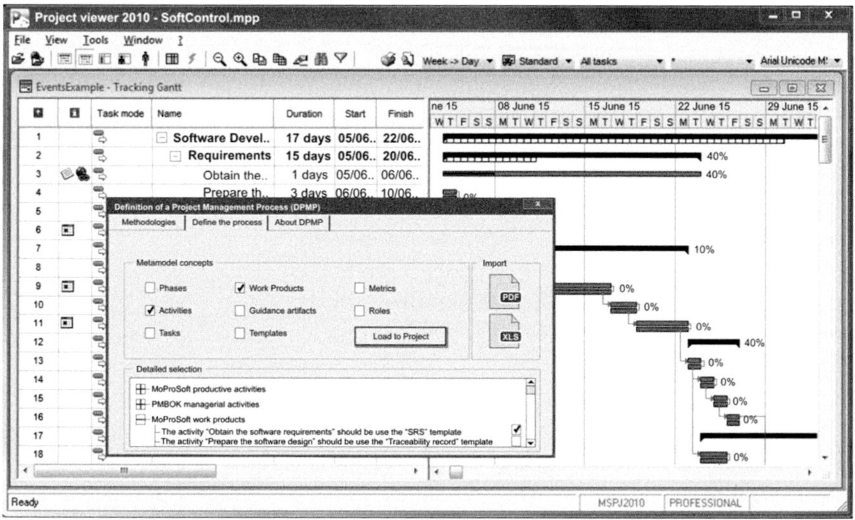

Fig. 4 The incorporation of the DPMP to Microsoft Project®

composed by a set of ***activities***. It is noteworthy that the architecture of our metamodel does not establish by default a set of strict steps or activities for project managers. According to the conceptual model presented above, when a request for a specific project is created, it is necessary that the project manager choose all the elements of his process. This process should take into the account the necessities and complexity of that project (e.g., Fig. 4 shows that an add-in had loaded to the project all the basic process elements of PMBOK® Guide, for managerial activities, and MoProSoft®, for productive activities). At the same time, activities can be broken down into more specific actions called ***tasks*** that require ***work products*** to facilitate their execution. Activities may have ***dependencies*** between them, which help to define the order in which they should be executed within the project. The activities are usually supported by some kind of ***guideline*** (e.g., a tool, technique, artifact, process asset) so that it is possible to verify the procedure to be performed before its execution. The activities typically produce ***outputs*** (e.g., deliverables, documents, artifacts), and depend on ***inputs*** to generate a result.

The ***roles*** in the metamodel are divided according to the type of activity to be performed (i.e., a management activity or a productive activity). Thus, each activity must be performed by one or more roles, typical of small teams. A similar relationship can occur with the ***resources*** (physical or human) because an activity may depend on certain infrastructure and knowledge to achieve the project's objective. Furthermore, any given activity has only one responsible (as it is indicated by the standards and methodologies for project management), so we try to eliminate the common problem of small-sized software enterprises that constantly change roles in each project. An activity can use a work product as input or output, but it can also modify an existing one; thus, this concept has been extended to allow three distinct associations: *generates*, *modifies*, and *uses* as inputs (i.e., an activity can generate a work product, or modify or use an existing one). It is important to mention that it has also been indicated the association "modifies" between the work product and the role, because it is the person who plays that role the one who has the possibility to change an artifact without performing any work. This consideration was also added to enable automation of an instantiated process considering the fact that it is not possible to eliminate an activity that creates a work product if there is another activity that modifies or uses as input the same product. Each work product must display information about its version (i.e., configuration management) and its type: "external" (when it is subject to the approval of a person or client) or "internal" (when is total or partially delivered to the role that is responsible of that activity). Finally, a work product and an activity should use ***measures*** to be verified and to determine its successful use, respectively.

Thus, the proposed metamodel provides a conceptual architecture that enables project managers to define a unique process to assist them in project planning and control monitoring taking into the account the concepts arising from the software development processes. To demonstrate the feasibility of the proposed idea, we have developed a software prototype called DPMP (Definition of a Project Management Process) as an add-in for Microsoft Project® 2007. Based on this idea, the concepts coming from the metamodel were added to this prototype (which has already included as start point the simplifications of PMBOK® Guide and MoProSoft® for managerial and productive activities, respectively). This choice allows small-sized software enterprises to take advantage of the features that are already implemented in Microsoft Project® with the proposed definition of a process for project management (see Fig. 4).

4 Conclusions

We started this paper by recognizing the common problem in project management that the small-sized software enterprises need to confront to support the success of a project. In this sense, many project management methodologies have been developed from industry practices and international standards to ensure a higher ratio of success for software projects. It is true that these methodologies have been widely and effectively used in large-sized software companies. However, when software projects are developed within the context of a small-sized software enterprise, there is often a lack of an appropriate method for project management or skilled project implementers who can use the methodologies used in large-sized organizations. Therefore, this research coincides with the findings of Sánchez-Gordón and O'Connor [30] in the sense that it is necessary to provide an insight toward a simplification of work products as they relate to software development process activities in small-sized enterprises for supporting the project success.

Thus, this paper has presented a proposal for simplifying the main concepts related to project management with a metamodel for defining a customized process in the context of small-sized software enterprises. Our study has initially identified the relevance of project management activities for these enterprises during a software project. Then, the lack of information about these skills in most small-sized software enterprises has been noticed. After an individual analysis of many methodologies for project management, we have proposed an alternative metamodel that aims to provide a simplified perspective of management. This metamodel identifies the basic elements that enable enterprises to focus their effort to improve at project-level and, in the future, at process-level. The definition of a customized process emphasizes the importance of individuals' skills to both develop and manage a software project and create a solid basis of knowledge to improve the software and product qualities.

Finally, we have created the DPMP add-in for Microsoft Project®, one of the most management tools used in the context of small-sized software enterprises, to evaluate the feasibility of our idea. We are currently collecting data through performing some case studies to provide more formal conclusions about the incorporation of the metamodel to define simplified processes in real projects and teams.

References

1. Turner, R., Ledwith, A., Kelly, J.: Project management in small to medium-sized enterprises: matching processes to the nature of the firm. Int. J. Project Manage. **28**(8), 744–755 (2010)
2. Richardson, I.: Why are small software organizations different? IEEE Softw. **24**(1), 18–22 (2007)
3. Ribaud, V., Saliou, P., O'Connor, R.V., Laporte, C.Y.: Software engineering support activities for very small entities. In: Riel, A., O'Connor, R., Tichkiewitch, S., Messnarz, R. (eds.) Systems, Software and Services Process Improvement, vol. 99, pp. 165–176. Springer, Heidelberg (2010)
4. Turner, R., Ledwith, A., Kelly, J.: Project management in small to medium-sized enterprises: tailoring the practices to the size of company. Manage. Decis. **50**(5), 942–957 (2012)
5. Henderson-Sellers, B.: Bridging metamodels and ontologies in software engineering. J. Syst. Softw. **84**(2), 301–313 (2011)
6. Callegari, D.A., Bastos, R.M.: Project management and software development processes: integrating RUP and PMBOK. In: 2007 International Conference on Systems Engineering and Modeling (ICSEM '07), pp. 1–8. IEEE Computer Society, California (2007)
7. Adolph, S., Kruchten, P., Hall, E.: Reconciling perspectives: a grounded theory of how people manage the process of software development. J. Syst. Softw. **85**(6), 1269–1286 (2012)
8. Feiler, P., Humphrey, W.: Software process development and enactment: concepts and definitions. Software Engineering Institute, Carnegie Mellon University, CMU/SEI-92-TR-004, Pittsburgh, Pennsylvania, USA (1992)
9. Clarke, P., O'Connor, R.V.: The situational factors that affect the software development process: towards a comprehensive reference framework. Inf. Softw. Technol. **54**(5), 433–447 (2012)
10. McCormack, A., Crandall, W., Henderson, P., Toft, P.: Do you need a new product-development strategy? Aligning process with context. Res. Technol. Manage. **55** (1), 34–43 (2012)

11. Spalek, S.: Does investment in project management pay off? Ind. Manage. & Data Syst. **114** (5), 832–856 (2014)
12. Ståhl, D., Bosch, J.: Modeling continuous integration practice differences in industry software development. J. Syst. Softw. **87**, 48–59 (2014)
13. Cerón, R., Dueñas, J.C., Serrano, E., Capilla, R.: A meta-model for requirements engineering in system family context for software process improvement using CMMI. In: Bomarius, F., Komi-Sirviö, S. (eds.) Product Focused Software Process Improvement, vol. 3547, pp. 173–188. Springer, Heidelberg (2005)
14. Méndez, D., Penzenstadler, B., Kuhrmann, M., Broy, M.: A meta model for artefact-orientation: fundamentals and lessons learned in requirements engineering. In: Petriu, D., Rouquette, N., Haugen, Ø. (eds.) Model Driven Engineering Languages and Systems, vol. 6395, pp. 183–197. Springer, Heidelberg (2010)
15. Goknil, A., Kurtev, I., van de Berg, K., Spijkerman, W.: Change impact analysis for requirements: A metamodeling approach. Inf. Softw. Technol. **56**(8), 950–972 (2014)
16. Henderson-Sellers, B., Gonzalez-Perez, C.: A comparison of four process metamodels and the creation of a new generic standard. Inf. Softw. Technol. **47**(1), 49–65 (2005)
17. Ayed, H., Vanderose, B., Habra, N.: A metamodel-based approach for customizing and assessing agile methods. In: 8th International Conference on the Quality of Information and Communications Technology (QUATIC), pp. 66–74, IEEE Computer Society, New York (2012)
18. García, F., Serrano, M., Cruz-Lemus, J., Ruiz, F., Piattini, M.: Managing software process measurement: A metamodel-based approach. Inf. Sci. **177**(12), 2570–2586 (2007)
19. Colombo, A., Damiani, E., Frati, F., Ontolina, S., Reed, K., Ruffatti, G.: The use of a meta-model to support multi-project process measurement. In: 15th Asia-Pacific Software Engineering Conference (APSEC '08), pp. 503–510. IEEE Computer Society, New York (2008)
20. Martins, P.V., da Silva, A.R.: PIT-ProcessM: a software process improvement meta-model. In: 7th International Conference on the Quality of Information and Communications Technology (QUATIC), pp. 453–458. IEEE Computer Society, New York (2010)
21. Tian, L., Zeng, G.Y., Yu, L., Zhu, B.: Research and implementation of software process metamodel for CMMI. Comput. Eng. Design **18**, 245–267 (2010)
22. Banhesse, E.L., Salviano, C.F., Jino, M.: Towards a metamodel for integrating multiple models for process improvement. In: 38th EUROMICRO Conference on Software Engineering and Advanced Applications (SEAA), pp. 315–318. IEE Computer Society, New York (2012)
23. Abels, S., Ahlemann, F., Hahn, A., Hausmann, K., Strickmann, J.: PROMONT—a project management ontology as a reference for virtual project organizations. In: Meersman, R., Tari, Z., Herrero, P. (eds.) On the Move to Meaningful Internet Systems, vol. 4277, pp. 813–823. Springer, Heidelberg (2006)
24. Velić, M., Padavić, I., Dobrović, Ž.: Metamodel of agile project management and the process of building with LEGO® bricks. In: 23rd Central European Conference on Information and Intelligent Systems (CECIIS), pp. 481–193. University of Zagreb, Varazdin (2012)
25. Mas, A., Mesquida, L.A.: Software project management in small and very small entities. In: 8th Iberian Conference on Information Systems and Technologies (CISTI), pp. 1–6. IEEE Computer Society, New York (2013)
26. Atkinson, C.: Metamodelling for distributed object environments. In: First International Enterprise Distributed Object Computing Workshop (EDOC'97), pp. 90–101. IEEE Computer Society, New York (1997)
27. International Organization for Standardization/International Electrotechnical Commission.: ISO/IEC 24744. Software Engineering—Metamodel for Development Methodologies. ISO, Geneva (2007)

28. Gonzalez-Perez, C., Henderson-Sellers, B.: Modelling software development methodologies: a conceptual foundation. J. Syst. Softw. **80**(11), 1778–1796 (2007)
29. Object Management Group. Software & Systems Process Engineering Meta-Model Specification, Version 2.0. http://doc.omg.org/formal/08-04-01.pdf (2008). Accessed 1 July 2015
30. Sanchéz-Gordón, M.L., O'Connor, R.V.: Understanding the gap between software process practices and actual practice in very small companies. Softw. Qual. J. (2015). doi:10.1007/s11219-015-9282-6

Can User Stories and Use Cases Be Used in Combination in a Same Project? A Systematic Review

Dennis Cohn-Muroy and José Antonio Pow-Sang

Abstract Requirement elicitation (RE) is one of the main tasks that must be performed in order to guarantee the correct implementation of a software development. Its incorrect specification can cause unnecessary overdue costs for the project and, in some cases, its complete failure. The objective of this paper is to provide a state of the art of the elicitation models that makes simultaneous use of two well-known techniques: the use cases model and user stories. The systematic literature review was chosen as a supportive investigation methodology. From the 45 found publications, the search strategy identified 11 studies and 3 methodological proposals: Athena, K-gileRE and NORMAP. Finally, after having reviewed the literature, it was found that there are a few validated proposals that makes use of the combination of user stories and use cases models. Also, there is not enough information to acknowledge the actual efficacy of combining both techniques.

Keywords NFR · Non functional requirements · Requirement elicitation · Requirements engineering · Software engineering · SLR · Systematic literature review · Use cases · User stories

1 Introduction

The requirement elicitation (RE) process is one of the main tasks to be performed in order to guarantee the correct implementation of a software product. Unnecessary costs or even the project failure can be derived from its incorrect implementation.

D. Cohn-Muroy (✉) · J.A. Pow-Sang
Escuela de Posgrado, Pontificia Universidad Católica del Perú, PUCP, Lima, Peru
e-mail: dennis.cohn@pucp.edu.pe

J.A. Pow-Sang
e-mail: japowsang@pucp.edu.pe

J. Mejia et al. (eds.), *Trends and Applications in Software Engineering*,
Advances in Intelligent Systems and Computing 405,
DOI 10.1007/978-3-319-26285-7_2

Boehm et al. [1] reported that the 45 % of errors that exists in a software product are originated during the RE and the preliminary design. Also, fixing these errors would demand more effort than fixing the errors originated during the coding process. In 1995, the Standish Group [2] presented in the Chaos Report that the ambiguity of the requirements and its incompleteness is one of the reasons that explains the failure of software projects.

According to Barbacci et al. [3] during the software RE, non functional requirements (NFR) related to the quality attributes of the software product tend to be ignored or ambiguously specified. Consequently, the software architect is the responsible on identifying and prioritizing these attributes. However, his decision may differ from the priorities and expectations of the stakeholders. Nord and Tomayko [4] indicated that in some software projects the architecture design task does not get enough importance.

There are two well-known techniques which may be used during the RE process. The first one is the use case modeling technique, which is a group of organized scenarios that are used to define the purpose of systems and software Alexander and Zink [5]. Its main goal is to help during the RE process defining the interactions between the environment (actors) and the system. In its description, the functionalities that the software must meet are included Gallardo-Valencia et al. [6].

Cockburn [7] defines use cases as a group of requirements that can detail the software behavior if they are correctly specified. Use cases do not require including all of the functionalities that had been indentified; however, they must include the most relevant ones. They have the following elements: (i) Name, (ii) Objective, (iii) Brief Description, (iv) Flow of events, (v) Preconditions, (vi) Post-conditions, (vii) NFR, (viii) Supportive Diagrams.

The second technique is known as user stories. According to Cohn [8] and Winbladh et al. [9], they are short stories that describe some feature that needs to be included in the system. They are centered in the needs of the user and are commonly used in agile projects. User stories have the following components: (i) The Card, which is a description of the story where the user role, the task to be performed and the task's goal are presented; (ii) The conversation, which contains additional information that can complement the Card; (iii) The confirmation, which is a group of tests that can be used to verify the completeness of the user story.

The objective of this paper is to provide an overview of the state of the art of the elicitation methods that simultaneously use the use cases model and the user stories. Then, identify how they can improve the RE process by adequately specifying NFR and which of them are not only proposals but have been validated by the academia or evaluated by the industry.

This document is organized as follows. Section 2 presents the review protocol that was followed; the proposed research questions; the execution of the search and the data analysis of the found results. Section 3 proposes answers for each one of the research questions based on the search results. In Sect. 4, the threats to validity of this study are analyzed. Finally, Sect. 5 presents the conclusions and proposes future studies that can be done.

2 Review Process

A Systematic Literature Review (SLR) was performed. According to Kitchenham and Charters [10], this method would allow the identification, evaluation and interpretation of a set of researches from the same topic. The objective of this study is the identification and review of RE modeling techniques that simultaneously use the user stories and the use cases model. Figure 1 shows the review protocol proposed by Ahmad et al. [11] that was followed on this paper.

2.1 Planning

Research Questions. The 5 PICOC (Population, Intervention, Comparison, Outcomes & Context) criteria were used in the construction of three secondary research questions that would help answering the main question:

RQ1: *Are there any proposed models that make simultaneous use of user stories and use cases model?*—This question makes use of the Population and Intervention (PI) categories of the PICOC technique. As shown in Table 1, the goal of this question is to obtain the current state of the art of all the available publications in software engineering or requirement engineering that makes simultaneous use of use cases and user stories.

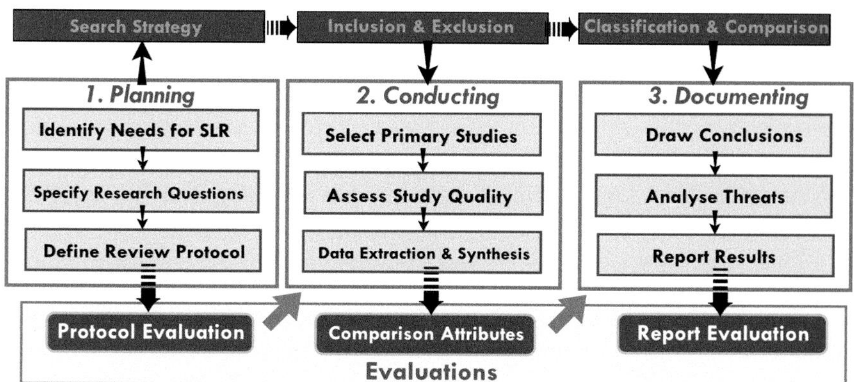

Fig. 1 Systematic literature review protocol proposed by Ahmad et al. [11]

Table 1 PICOC for Q1

Criteria	Scope
Population	Software engineering OR software requirements OR software development
Intervention	Techniques that make use of uses cases and user stories

Table 2 PICOC for Q2

Criteria	Scope
Population	Software engineering OR software requirements OR software development
Intervention	Techniques that make use of uses cases and user stories
Outcome	Non functional requirements elicitation

Table 3 PICOC for Q3

Criteria	Scope
Population	Software engineering OR software requirements OR software development
Intervention	Techniques that make use of uses cases and user stories
Context	Application on academia or industry

RQ2: *Do any of the proposed models take into account the elicitation of non functional requirements?*—Making use of the results achieved through RQ1, this question's goal is to find which of the proposed models considers the elicitation of NFR. Table 2 shows the structure of the question.
RQ3: *Do any of the proposed models have been validated or are used by the industries?*—From the results of the RQ1 and considering the paper classification proposed by Wieringa et al. [12]: evaluation, proposal, validation, philosophy, opinion, personal experiences. This question's goal is to find out if any of the proposed models has been applied in the industry or in the academia. Table 3 shows the structure of the question.

2.2 *Execution*

Selection of Studies. The following inclusion and exclusion criteria were applied over the publications found during the primary and secondary searches.

Inclusion Criteria. (i) The publications must be written in English. (ii) The full text of the paper must be accessible. (iii) Only the following classifications will be considered: reviews, proposals, validations, evaluations [12].

Exclusion Criteria. (i) Papers which are not proposing a model that makes simultaneous use of use cases and user stories. (ii) Papers which are not validating or evaluating the combined use of both techniques, use cases and user stories. (iii) In the case of duplicated papers, only the most complete paper will be considered.

Search Strategy. For the primary search, the 5 steps proposed by Ahmad et al. [11] were followed. These steps are described in Table 4.

For each one of the research questions, a string query was built:

SQ1: ((“user story” OR “user stories”) AND (“use case” OR “use cases”)) AND (“software development” OR “software construction” OR “software project” OR

Table 4 Steps performed during the primary search

Steps	Description
Build the search queries	The search queries were built based on the terms listed on the Tables 1, 2 and 3
Consider synonyms	Synonyms were considered for each term used on the search
Combine the search terms	The logical connector "OR" was used to connect the synonyms and the connector "AND" was used for connecting the criteria
Divide the search string	The search string was divided in substrings so they can be executed on the different data sources
Manage the found references	The Mendeley tool was used for managing the references

"software projects" OR "software process" OR "software processes" OR "software engineering" OR "requirement* engineering" OR "requirement*")
SQ2: **SQ1** AND ("Non Functional Requirement*" OR NFR)
SQ3: **SQ1** AND (study OR studies OR experiment* OR verificat* OR validat* OR evaluation*)

The queries were executed on April 2015 in the following data sources: Sciverse Scopus (SS, http://scopus.com/), IEEExplore (IEEE, http://ieeexplore.ieee.org), ACM Digital Library (ACM, http://dl.acm.org/) and ISI Web of Science (ISI, http://isiknowledge.com). The included search fields were title, abstract and keywords only if the data source provided those options. No publication year filter was applied over the search.

The primary search returned a total of 42 studies published between the years 2003 and 2014. There were 16 duplicated studies. Then, the inclusion and exclusion criteria were applied over the abstract of the unique results reducing the number of papers to 15. Finally, the full text of the 15 papers was reviewed, reducing the number of relevant studies to 8.

On the secondary search there were reviewed (i) papers that cite the primary search results and (ii) papers that are cited by the primary search results. In both cases the selected papers must fulfill with the inclusion and exclusion criteria. Through applying the secondary search to the abstract of the studies, 3 additional publications were found.

The full list of selected publications is shown on Table 5.

Studio Quality Assessment. The quality of the 11 selected studies was evaluated using the checklist proposed by Zarour et al. [23]. For each question, there were 3 possible answers (Yes = 1 point; No = 0 point; Partially = 0.5 point). The quality assessment checklist and the results are presented on Tables 6 and 7.

Only publications with a score higher than 2.5 were accepted. From the selected studies, the lowest score achieved was 3.0 and the average was 3.95. Then, the result of the quality assessment test shows that the 11 publications are acceptable.

Data Extraction and Synthesis. The next step was to extract and synthesize the data of the selected publications listed on Table 4.

Table 5 Selected studies from the primary and secondary search

Reference	Digital Source	Proposal	Classification [12]	NFR	Type
Farid [13]	–	NORMAP	Proposal validation	Yes	Secondary
Farid [14]	ACM	NORMAP	Proposal	Yes	Primary
Farid et al. [15]	SS, IEEE, ISI	NORMAP	Proposal validation	Yes	Primary
Farid et al. [16]	SS, IEEE, ISI	NORMAP	Proposal validation	Yes	Primary
Farid et al. [17]	–	NORMAP	Proposal validation	Yes	Secondary
Gallardo-Valencia [6]	SS, IEEE, ACM	–	Validation	No	Primary
Gallardo-Valencia [18]	SS	–	Validation	No	Primary
Hvalshagen [19]	SS	–	Philosophy validation	No	Primary
Kumar et al. [20]	SS, ACM	K-gileRE	Proposal validation	No	Primary
Laporti et al. [21]	SS, ACM, ISI	Athena	Proposal Validation	Yes	Primary
Liskin [22]	–	–	Validation evaluation	No	Secondary

Table 6 Quality assessment checklist

ID	Question	Yes	Partially	No
QA1	Is the aim of the research sufficiently explained?	11	0	0
QA2	Is the presented idea/approach clearly explained?	10	1	0
QA3	Are threats to validity taken into consideration?	2	2	7
QA4	Is there an adequate description of the context in which the research was carried out?	10	0	1
QA5	Are the findings of the research clearly stated?	7	4	0

Publications yearly distribution. From the results, it was found that all of the selected studies were published from 2007 to 2015. The years 2007 and 2012 were the ones with the most studies published. However, there were no published studies on 2014.

Most relevant publishing sources. According to the information on Table 5, IEEE SOUTHEASTCON is the source where the most number of publications has been published. Also, 81.8 % of the publications were Conference papers, 9.1 % were Journal papers (Computers in Industry) and the rest were Thesis publications (NSU Thesis).

Table 7 Quality assesment result per selected publication

References	QA1	QA2	QA3	QA4	QA5	Score
Farid [13]	1.0	0.5	0.0	1.0	1.0	3.5
Farid [14]	1.0	1.0	0.0	1.0	1.0	4.0
Farid et al. [15]	1.0	1.0	0.0	1.0	1.0	4.0
Farid et al. [16]	1.0	1.0	0.0	1.0	1.0	4.0
Farid et al. [17]	1.0	1.0	0.0	1.0	0.5	3.5
Gallardo-Valencia et al. [6]	1.0	1.0	1.0	1.0	1.0	5.0
Gallardo-Valencia et al. [18]	1.0	1.0	0.0	1.0	0.5	3.5
Hvalshagen et al. [19]	1.0	1.0	0.0	0.0	1.0	3.0
Kumar et al. [20]	1.0	1.0	0.5	1.0	0.5	4.0
Laporti et al. [21]	1.0	1.0	0.5	1.0	0.5	4.0
Liskin [22]	1.0	1.0	1.0	1.0	1.0	5.0

Publications' synthesis. After reviewing the full text of the selected studies, there were identified 3 RE proposals that make simultaneous use of use cases models and user stories (Athena, K-gileRE and NORMAP). The main characteristics of each proposal were extracted in order to compare their similarities and differences. In all of them, the user stories are applied differently before the use cases. Athena and K-gileRE construct the user stories through collective communication between the stakeholders and analysts. In NORMAP, each of the system requirements must be converted into a user story following the W^8 card format extension.

Furthermore, although Athena mentions its capacity for handling NFR elicitation, that skill depends on how much detail has been included in the user stories. Consequently, the method will fail if the information related to the NFR requirements is ambiguous or incomplete. On the other hand, NORMAP has pre-established a list of metrics that allows the correct detection of NFR and its impact during the project's risk analysis.

3 Discussion

This Section presents the answers for each of the research questions proposed on Sect. 2.

RQ1: *Are there any proposed models that make simultaneous use of user stories and use cases model?*

Three RE methods that combines the use of user stories and use cases have been identified from the selected papers: Athena [21], K-gileRE [20] and NORMAP [13–17].

RQ2: *Do any of the proposed models take into account the elicitation of non functional requirements?*

From the found methods, NORMAP makes use of an user story extension for getting additional information of the client necessities like quality attributes prioritization and impact analysis evaluation. It also proposes 3 new components for the requirements modeling: AUC (a simplified use case diagram), ALC (for representing NFR associated to an AUC) and ACC (proposed solutions for an ALC).

On the other hand, even though the authors of Athena affirm that it can be used as a tool for NFR elicitation, this method does not have enough tools to guarantee the correct elicitation of these kinds of requirements.

RQ3: *Do any of the proposed models have been validated or are used by the industries?*

From the 3 found methods, none of them has been evaluated in a non-academic environment. In the case of Athena, its authors proposed to evaluate the method in a real environment in a future study. K-gileRE has only been validated in a small project where it showed good results; but it is necessary to validate the method in bigger projects, so its behavior in real projects can be predicted. Moreover, the NORMAP method was validated making use of a real project data and documentation; however it is necessary to validate the method over a more diverse list of projects before its results can be generalized.

Even though the papers of Hvalshagen et al. [19] and Gallardo-Valencia et al. [6, 18] do not propose a RE methodology, the results from their experiments related to the combined use of user stories and use cases bring a significant input for the proposal of new solutions or for improving any of the found techniques.

4 Threats to Validity

According to Jedlitschka et al. [24] there are 4 possible threads to validity that must be discussed.

Construct Validity: The search queries were built from words' synonyms of each research question's PICOC and were ran in each of the selected data sources. Some relevant studies might have not been included due to being indexed in non-included data sources or they might have terms which were not considered during the building of the query.

Internal Validity: In order to mitigate the risk of the study selection bias, the adviser validated the analysis performed by the main researcher.

External Validity: In order to mitigate the threat related to the incapability of generalizing the results of this study, the search process was run multiple times.

Conclusion Validity: In order to mitigate the threat of excluding relevant studies during the SLR, the inclusion and exclusion criteria were carefully built. Also, the quality of the relevant studies was evaluated through a quality assessment checklist.

5 Conclusion and Future Work

The objective of this review was to identify and evaluate RE methods that simultaneously used user stories and use cases modeling. From the 45 found studies, 11 publications were selected. From them, 3 proposal solutions were found: Athena, K-gileRE and NORMAP. Each of these proposals has been validated in an academic controlled environment; however, none of them have been evaluated in a real life environment. Consequently, it is not possible to measure their actual efficacy.

Moreover, each of the methods has some characteristics and assumptions that do not necessarily apply to a real environment situation. In Athena and K-gileRE there is no guarantee that the stakeholders will willingly have the availability and interest to work collectively or that the analysts will share with the stakeholders their knowledge on the requirements analysis. In the case of NORMAP, because of its complexity, is highly dependable of its CASE tool (NORMATIC) even though it was developed for being applied in agile projects. Furthermore, in spite of the fact that its proposed use cases give input related to the NFR of the software to be implemented, it does not include the necessary information to identify which actor initializes a specific use case.

In conclusion, there have been a few publications that has presented the advantages (or disadvantages) of combining the use of both techniques, so future studies on this topic can analyze the benefits of applying these techniques on agile (or non-agile) projects and how its combination can affect the level of ambiguity or completeness of requirements during the software development life cycle. Also, from the identified solution proposals, it is recommended to evaluate them in the industry in order to measure their efficacy and how much they do satisfy the needs of the people who applied them during the elicitation process.

The next step will be to evaluate if the simultaneous usage of user stories and use cases benefits communicability and transparency during the requirements elicitation.

References

1. Boehm, B.W., McClean, R.K., Urfrig, D.B.: Some experience with automated aids to the design of large-scale reliable software. ACM SIGPLAN Not **10**, 105–113 (1975)
2. Standish Group: CHAOS Report (1995)
3. Barbacci, M.R., Ellison, R., Lattanze, A.J., Stafford, J. a., Weinstock, C.B., Wood, W.G.: Quality Attribute Workshops, 3rd edn. (2003)
4. Nord, R.L., Tomayko, J.E.: Software architecture-centric methods and agile development. IEEE Softw. **23**, 47–53 (2006)
5. Alexander, I., Zink, T.: Introduction to systems engineering with use cases. Comput. Control Eng. J. **13**, 289–297 (2002)
6. Gallardo-Valencia, R.E., Olivera, V., Sim, S.E.: Are use cases beneficial for developers using agile requirements? In: 2007 Fifth International Workshop on Comparative Evaluation in Requirements Engineering, pp. 11–22. IEEE (2007)

7. Cockburn, A.: Writing effective use cases (2001)
8. Cohn, M.: User Stories Applied: For Agile Software Development. Addison Wesley Longman Publishing Co., Inc, Redwood City (2004)
9. Winbladh, K., Ziv, H., Richardson, D.J.: Surveying the usability of requirements approaches using a 3-dimensional framework (2008)
10. Kitchenham, B., Charters, S.: Guidelines for performing systematic literature reviews in software engineering. Engineering **2**, 1051 (2007)
11. Ahmad, A., Jamshidi, P., Pahl, C.: Protocol for Systematic Literature Review (2012)
12. Wieringa, R., Maiden, N., Mead, N., Rolland, C.: Requirements engineering paper classification and evaluation criteria: a proposal and a discussion. Requir. Eng. **11**, 102–107 (2006)
13. Farid, W.M.: The Normap methodology: lightweight engineering of non-functional requirements for agile processes. In: Proceedings—Asia-Pacific Software Engineering Conference, APSEC, pp. 322–325 (2012)
14. Farid, W.M.: The Normap Methodology: Non-functional Requirements Modeling for Agile Processes (2011)
15. Farid, W.M., Mitropoulos, F.J.: NORMATIC: a visual tool for modeling non-functional requirements in agile processes. In: 2012 Proceedings of IEEE Southeastcon, pp. 1–8. IEEE (2012)
16. Farid, W.M., Mitropoulos, F.J.: Novel lightweight engineering artifacts for modeling non-functional requirements in agile processes. In: 2012 Proceedings of IEEE Southeastcon, pp. 1–7. IEEE (2012)
17. Farid, W.M., Mitropoulos, F.J.: Visualization and scheduling of non-functional requirements for agile processes. In: Conference Proceedings—IEEE SOUTHEASTCON (2013)
18. Gallardo-Valencia, R.E., Olivera, V., Sim, S.E.: Practical experiments are informative, but never perfect. In: Proceedings of the 1st ACM International Workshop on Empirical Assessment of Software Engineering Languages and Technologies Held in Conjunction with the 22nd IEEE/ACM International Conference on Automated Software Engineering (ASE) 2007—WEASELTech '07, pp. 37–42. ACM Press, New York, NY, USA (2007)
19. Hvalshagen, M., Khatri, V., Venkataraman, R.: Understanding use cases: Harnessing the power of narratives to comprehend application domains. In: 14th Americas Conference on Information Systems, AMCIS 2008, pp. 3328–3335 (2008)
20. Kumar, M., Ajmeri, N., Ghaisas, S.: Towards knowledge assisted agile requirements evolution. In: Proceedings of the 2nd International Workshop on Recommendation Systems for Software Engineering—RSSE '10, pp. 16–20. ACM Press, New York, NY, USA (2010)
21. Laporti, V., Borges, M.R.S., Braganholo, V.: Athena: A collaborative approach to requirements elicitation. Comput. Ind. **60**, 367–380 (2009)
22. Liskin, O.: How artifacts support and impede requirements communication. In: 21st International Working Conference, REFSQ 2015, pp. 132–147 (2015)
23. Zarour, M., Abran, A., Desharnais, J.-M., Alarifi, A.: An investigation into the best practices for the successful design and implementation of lightweight software process assessment methods: a systematic literature review. J. Syst. Softw. **101**, 180–192 (2015)
24. Jedlitschka, A., Ciolkowski, M., Pfahl, D.: Reporting experiments in software engineering. In: Guide to Advanced Empirical Software Engineering, pp. 201–228 (2008)

Addressing Product Quality Characteristics Using the ISO/IEC 29110

Gabriel Alberto García-Mireles

Abstract Several software process models have been proposed to increase the competitiveness of small companies based on the idea that increasing the capability of software process can increase the quality of the software product. Software quality is a complex concept which can be studied from diverse perspectives, such as process and product. The aim of this work is to identify to what extent product quality characteristics are addressed in process models oriented towards very small companies. A mapping method is used to study the ISO/IEC 29110 basic profile in order to select practices directly related to product quality. The quality characteristics under study are described in ISO/IEC 25010. The results show that correctness and testability are mandatory quality characteristics, whereas usability is optional. However, they are only addressed during analysis activities. This result can lead to the development of new profiles oriented towards increasing product quality throughout the software life-cycle.

Keywords Product quality · Very small organizations · Software process

1 Introduction

Software quality is a main driver that organizations, including small ones, need to take into account in order to survive and grow. A common approach organizations use is to implement software quality models. However, small software companies have different requirements when implementing them [1].

Although several research groups have developed models for software process improvement and assessment in small companies, subcommittee 7 of ISO/IEC started a project to address software lifecycle processes for very small companies

G.A. García-Mireles (✉)
Departamento de Matemáticas, Universidad de Sonora, Blvrd. Encinas y Rosales s/n, 83000 Hermosillo, Sonora, Mexico
e-mail: mireles@mat.uson.mx

J. Mejia et al. (eds.), *Trends and Applications in Software Engineering*, Advances in Intelligent Systems and Computing 405,
DOI 10.1007/978-3-319-26285-7_3

[2]. As a result of this effort, the ISO/IEC 29110 was developed. Its goal is to support very small organizations in assessing and improving their software process. A very small entity (VSE) is defined as an "entity (enterprise, organization, department or project) having up to 25 people" [3].

VSEs are an important sector that significantly contribute to economic growth in many countries [4]. VSEs involved in software development can provide software technology for the domestic market as well as the offshore market. In addition, they can become suppliers of software components in the manufacturing sector. In this setting, a defective software component integrated in a product can have expensive consequences [5]. Therefore, VSEs have a pressing need to develop their products efficiently, effectively and with high quality [1].

Software quality can be viewed from distinct perspectives, for instance, considering process view and product view. In the former a set of practices is recommended to eventually optimize the software process. The latter view takes into account product characteristics that meet users' needs [6]. Product quality characteristics relevant to software quality are described in models such as ISO/IEC 25010 [7]. A holistic view of quality should address both quality perspectives—process and product [6]. Taking into account that one purpose of implementing process models in VSEs is to enhance product quality, in this work a report of the extent to which ISO/IEC 29110 Basic Profile [8] address product quality is presented. The identification of the activities within processes can support the development of improvement strategies to enhance product quality.

In order to identify those activities that can contribute to enhancing product quality, a mapping between the product quality model (ISO/IEC 25010) and the process model (ISO/IEC 29110 Basic Profile), was performed. Mappings between quality approaches are considered an effective way to identify requirements from different quality models and are useful to support multimodel appraisals [9]. As a result, the outcome of the mapping can be the starting point for developing a software process initiative that considers both product and process quality perspectives.

The remainder of the paper is organized as follows: Sect. 2 describes related work, and Sect. 3 gives a description of the mapping method. The results obtained are presented in Sect. 4. Section 5 is the discussion of results, while the conclusions and future work are addressed in Sect. 6.

2 Related Work

Quality approaches can be categorized based on the main artifacts targeted: product or process. It is recognized that software organizations must deal with several quality models simultaneously [9, 10]. However, several issues can arise when an organization implements them [11]. The differences among quality models with regard to their structure, granularity, and vocabulary can erode the expected benefits from an improvement initiative [9]. Furthermore, the majority of mappings between quality models have been carried out with process models [9, 12]. Few reports have described how to integrate both process and product quality views.

Several mappings between process and product quality standards have been reported in the literature. Ashrafi [13] suggested that a particular process model, CMM or ISO 9000, contributed towards enhancing a particular set of product quality characteristics. A similar approach was taken by Pardo et al. [14]. They provided a decision tree to show the influence of CMMI-Dev 1.2 and ISO 90003 on the ISO/IEC 25010 product quality model. In addition, García-Mireles et al. [15] carried out a harmonization between ISO 25010 and process models, such as ISO/IEC 12207 and CMMI. They found that product quality characteristics are described in activities suggested for elicitation and analysis of requirements.

Finally, García-Mireles et al. [6] found that the majority of mappings between process and product models have been performed for security related process models. Other mappings have focused on reliability, maintainability and security and have been performed on process models such as ISO/IEC 12207 or CMMI. However, the report by García-Mireles et al. does not include process models specifically targeted to VSEs.

3 Mapping Method

The method presented herein was derived from the method presented in [15]. The main difference is that the method applied in this paper includes the formal identification of the elements under comparison. This change correspond to one of the practices for preparing standardized profiles [16]. The activities included in the mapping method are:

1. Analyze models. The purpose of this activity is to identify the goals of the quality models, describe their structure and requirements.
2. Design mapping. The purpose of this activity is to establish a procedure for carrying out the mapping.
3. Execute mapping. The purpose of this activity is to perform the mapping between quality models.
4. Prepare a report. The purpose is to present a report with the results of the mapping between models.

4 Mapping ISO/IEC 25010 onto ISO/IEC 29110

The result of applying the mapping between the process standard (ISO/IEC 29110) and the product quality model described in ISO/IEC 25010 is presented in this Section, which is organized according to the activities described above.

4.1 Analyze Models

ISO/IEC 25010 [7] is a standard which describes the quality in use model and the product quality model. The quality in use model is composed of five characteristics, which are named effectiveness, efficiency, satisfaction, freedom from risk, and context coverage. Its primary concern is quality when software is used in the operation stage of its life cycle. The product quality model identifies eight quality characteristics, which are named functional suitability, performance efficiency, compatibility, usability, reliability, security, maintainability, and portability. It addresses quality when software is in the development stage.

Figure 1 depicts the quality characteristics belonging to the product quality model. The number inside the parenthesis shows the number of subcharacteristics included within each of the quality characteristics. For instance, the maintainability quality characteristic contains five subcharacteristics: modularity, reusability, analyzability, modifiability, and testability. In this work I have focused on the product quality model, since my interest is in addressing product quality during software development.

The ISO/IEC 25010 provides consistent terms and definitions to address relevant quality characteristics of all software products [7]. The quality models can be used to specify, measure, and evaluate software quality. Thus, ISO/IEC 25010 can be applied in different stages of the software lifecycle, including in the requirements, development, use, and maintenance stages.

The ISO/IEC 29110 [3] includes a set of standards and technical reports targeted to very small organizations, with the aim of improving both software quality and process performance. The expected benefits include an improved internal management process, enhanced customer satisfaction, improved product quality, and a decrease in development costs. Taking into account the limited resources of very small enterprises, the ISO/IEC 29110 focuses on the project management process and software implementation process. These processes were derived from the ISO/IEC 12207 software lifecycle processes standard [17]. The products described inside these processes rely on the ISO/IEC 15289 information products [8].

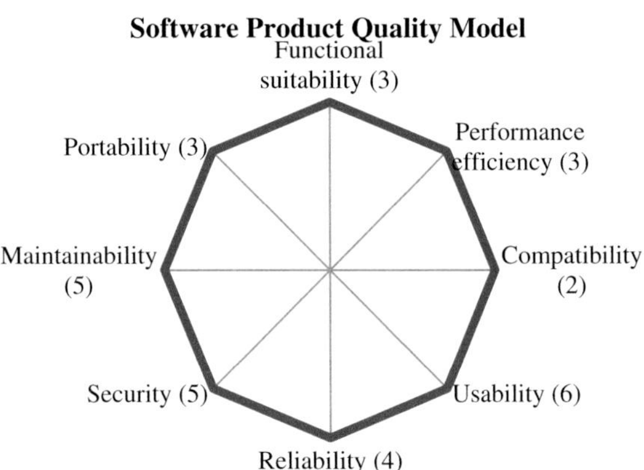

Fig. 1 Quality characteristics from ISO/IEC 25010

A primary component of the ISO/IEC 29110 is the profile that a VSE can apply to implement specific practices, which they do by means of guidelines published as technical reports. A profile is a "set of one or more base standards and/or standardized profiles and, where applicable, the identification of chosen classes, conforming subsets, options and parameters of those base standards, or standardized profiles necessary to accomplish a particular function" [3]. A profile group is a "collection of profiles which are related either by composition of processes (i.e. activities, tasks) or by capability level, or both" [3]. Each profile must be related to additional components such as an assessment guide and at least one implementation guide. Deployment packages can be developed as optional material to facilitate the implementation of profiles [3].

ISO/IEC 29110 includes the generic profile group which is applicable to VSEs that do not develop critical software products [16]. The profile group is decomposed into initial, basic, intermediate, and advanced. Currently, the basic profile is published as a technical report [8]. In this work I focus on the Basic Profile ISO/IEC 29110 to determine the best approach for addressing product quality. In order to meet a conformance profile, a VSE must demonstrate that (1) the work product developed is conformant with content described in the mandatory information products and that (2) the current practices applied in a software project produce the mandatory products described in the profile processes. Process, activities, objectives, work products, and outputs are mandatory. Tasks and inputs are optional.

The basic profile consists of two processes, the project management process and the software implementation process [8]. The purpose of the former process is to establish, and perform systematically, software development activities with the expected quality and within time and cost restrictions. The purpose of the latter process is to systematically carry out the activities of analysis, design, construction, integration and testing of a software product that meets the specified requirements [8]. Both processes are described in terms of purpose, objectives, input products, output products, internal products, roles involved, the flow of information diagram, activities, and tasks.

4.2 Design the Mapping

The purpose of the mapping is to identify those activities in a process model that address product quality characteristics. Thus, ISO/IEC 25010 was used as a source for product quality vocabulary, terms and definitions [7]. From the ISO/IEC 29110 [8] each process, goal, purpose, process, activity, task and work product was compared with the product quality characteristics. The method used tables to identify each element from the ISO/IEC 29110 standard. Each table row contains the description of a process element. If a process element mentioned a product quality characteristic then the row was linked to the appropriate clause from ISO/IEC 25010.

4.3 Execute the Mapping

The mapping was executed after the design mapping activity. In order to preserve the relationship between both standards, an excel worksheet was used for each process element from. For instance, the mapping between process objectives from the Software Implementation Process and product quality characteristics is presented in Table 1. The template contains columns that allow the identify cation of each element, as they are described in the standards, and includes a column to verify if the element description addresses any product quality characteristic. In this case, only the objective SI.O2 mentions two quality sub-characteristics: correctness and testability. The former belongs to the functional suitability characteristic while the latter to the maintainability quality characteristic. Process objectives from the Project Management Process do not address product quality characteristics; hence, this process is no longer used in this mapping.

The comparison of activities described in ISO/IEC 29110 used the same structure of Table 1. We found that activity SI.2 Software requirements analysis (from ISO/IEC 25010) is related to objective SI.O2 (from ISO/IEC 29110) and this activity also mentions correctness and testability as quality sub-characteristics.

When tasks were compared, we found that two of the seven tasks related to SI.2 software requirements analysis activity (from ISO/IEC 25010) mention quality

Table 1 Mapping objectives from the software implementation process (from ISO/IEC 29110) with product quality characteristics (from ISO/IEC 25010)

Software implementation objective	Brief description	Product quality?	Product quality characteristic—Reference ISO/IEC 25010
SI.O1	Tasks are performed as regards the project plan	0	
SI.O2	Software requirements are defined and analyzed for **correctness and testability**, …	1	4.2.1.2 functional correctness. 4.2.7.5 testability
SI.O3	Software architectural and detailed design is developed and baselined	0	
SI.O4	Software components defined by the design are produced	0	
SI.O5	Software components are integrated and verified	0	
SI.O6	Software configuration is integrated and stores at project repository	0	
SI.O7	Verification and validation tasks are performed	0	
	Total	**1**	

characteristics (from ISO/IEC 29110). The task SI.2.3 suggests verifying the correctness and testability of the requirements specification. This task is consistent with the objective SI.O2. The task SI.2.4 suggests "validate that requirements specification satisfies needs and agreed upon expectations, including the user interface usability" [11]. However, usability is not explicitly mentioned in process objectives and activities.

With regard to work products, we found that requirements specification addresses the majority of quality characteristics described into ISO/IEC 25010, except security and compatibility. Software design document suggests addressing performance efficiency, security, usability and reliability. However, there is an inconsistency, since security is not mentioned as a content in the requirements document but it is included in design. The software user documentation and maintenance documentation ask for content understandability. This quality subcharacteristic corresponds to usability.

4.4 Prepare a Report

The analysis of mapping between process reference and product quality model showed that ISO/IEC 29110 is aware of the need to improve product quality. From the process perspective, only Objective 2 (SI.O2) of the Software Implementation Process addresses two quality sub-characteristics, named as correctness and testability. These subcharacteristics must be defined and analyzed during the software requirements analysis activity (SI.2), since objectives and activities are mandatory parts of the process, as described in ISO/IEC 29110. The task SI.2.3 suggests verifying and obtaining approval of requirements using a verification approach. In addition, task SI.2.4 recommends validating the usability of the user interface. Although usability is not explicitly addressed in the process objectives, it is important to recognize its relevance for products developed by VSEs.

When focusing on the work products included in the Software Implementation Process, we found several work products that address product quality characteristics: requirements specification, software design, the product operation guide, software user documentation, and the maintenance documentation. However, quality characteristics addressed in the requirements specification are very difficult to trace in other work products of the software implementation process. In addition, there is a lack of definition of terms related to quality characteristics.

The requirements specification addresses several quality characteristics and subcharacteristics as they are described in the ISO/IEC 25010 (Fig. 2). These quality characteristics are addressed as quality requirements. It is important to note that the security and compatibility quality characteristics are not addressed in the description of requirements specification document [8]. This inconsistency could provide a basis to harmonize the ISO/IEC 15289 with ISO/IEC 25010.

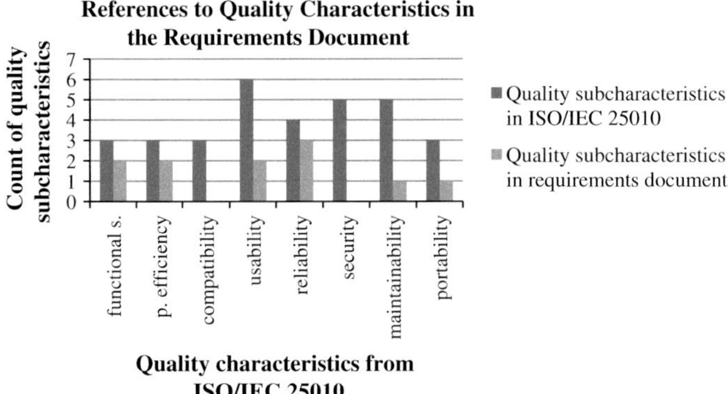

Fig. 2 Extent to which quality subcharacteristics in ISO/IEC 25010 are addressed in the requirements specification document

5 Discussion

The implementation of process models focused on the characteristics of small companies can improve software quality. This goal can be achieved by implementing the set of practices suggested in the ISO/IEC 29110. In addition, this standard shows that small companies also need to be aware of product quality characteristics. It is necessary to define and verify the correctness of the software specification and testability characteristics. However, the standard does not provide a definition of each product quality characteristic. Thus, ISO/IEC 25010 can be used to establish a baseline vocabulary for managing software product quality.

From the two processes described in ISO/IEC 29110, only the software implementation process refers explicitly to product quality characteristics. However, the mandatory process elements only address quality requirements during the validation activities of the requirements document. When the requirements document is analyzed, other quality characteristics are mentioned as optional. In addition, the design document, the maintenance document, and the user documentation address several optional quality characteristics.

In order to support VSEs in their effort to enhance software product quality, it is necessary to provide guidelines to address practices that improve quality characteristics. The guidelines should state how the quality characteristics are addressed in each stage of the software development lifecycle. In order to improve usability, for example, there are several methods that can be applied in different stages of software development [18, 19]. Thus, in order to provide appropriate support for a specific quality characteristic, the software organizations need to take into account additional practices. As a result, it is necessary to provide reference models, or profiles, that can help VSEs to achieve their specific product quality goals.

6 Conclusions

A mapping between the process and product quality models is presented. The results show that the ISO/IEC 29110 Basic Profile includes very few references to product quality characteristics as they are described in ISO/IEC 25010. The mandatory quality subcharacteristics of testability and correctness are addressed during verification activities of a software requirements document. Another quality characteristic, usability, is considered optional in the profile. When work product content is reviewed, several other quality characteristics and subcharacteristics are addressed as optional elements.

Beyond the software requirements verification activities identified to deal with product quality, the mapping results also suggest future work that is needed. Correctness and testability can be verified if they were previously taken into account when the work product was developed. Thus, it is necessary to consider those activities that aim to enhance a particular quality characteristic.

By considering the difference between product and process perspectives the consistency between ISO/IEC 29110 profiles and quality characteristics that must be considered in each outcome product can be seen. VSEs require guidelines to introduce practices that are oriented towards enhancing specific software quality goals. In addition, the mapping outcomes presented in this paper should be validated in future work by conducting empirical studies. Finally, the results of this work have the potential to lead to the development of new profiles oriented towards increasing product quality throughout the software life-cycle.

References

1. Basri, S., O'Connor, R.V.: Understanding the perception of very small software companies towards the adoption of process standards. In: Riel, A., O'Connor, R.V., Tichkiewitch, S., Messnarz, R. (eds.) EuroSPI 2010 CCIS, vol. 99, pp. 153–164. Springer, Berlin (2010)
2. Laporte, C.Y., O'Connor, R.V.: Systems and software engineering standards for very small entities: Implementation and initial results. In: 9th International Conference on the Quality of Information and Communications Technology, QUATIC 2014, pp. 38–47. IEEE, Guimaraes (2014)
3. ISO: ISO/IEC 29110-1:2011 Software engineering—lifecycle profiles for Very Small Entitites (VSEs)—part 1: overview (2011)
4. Muñoz, M., Gasca, G., Valtierra, C.: Caracterizando las necesidades de las pymes para implementar mejoras de procesos software: una comparativa entre la teoría y la realidad. Rev. Ibér. Sistemas y Tecnologías Información **E1**, 1–15 (2014)
5. Laporte, C.Y., O'Connor, R.V.: A systems process lifecycle standard for very small entities: development and pilot trials. In: Barafort, B., O'Connor, R.V., Poth, A., Messnarz, R. (eds.) EuroSPI 2014. CCIS, vol. 425, pp. 13–24. Springer, Berlin (2014)
6. García-Mireles, G.A., Moraga, M.Á., García, F., Piattini, M.: Approaches to promote product quality within software process improvement initiatives: a mapping study. J. Syst. Softw. **103**, 150–166 (2015)

7. ISO: ISO/IEC FCD 25010: Systems and software engineering—system and software product quality requirements and evaluation (SQauRE)—system and software quality models (2010)
8. ISO: ISO/IEC TR 29110-5-1-2:2011 Software engineering—lifecycle profiles for Very Small Entities (VSEs)—part 5-1-2: Management and engineering guide: Generic profile group: Basic profile (2011)
9. Kelemen, Z.D., Kusters, R., Trienekens, J.: Identifying criteria for multimodel software process improvement solutions-based on a review of current problems and initiatives. J. Softw. Evol. Process. **24**, 895–909 (2012)
10. Pardo, C., Pino, F.J., Garcia, F., Baldassarre, M.T., Piattini, M.: From chaos to the systematic harmonization of multiple reference models: a harmonization framework applied in two case studies. J. Syst. Softw. **86**, 125–143 (2013)
11. Henderson-Sellers, B.: Standards harmonization: theory and practice. Softw. Syst. Model. **11**, 153–161 (2012)
12. Pardo, C., Pino, F.J., García, F., Velthius, M.P., Baldassarre, M.T.: Trends in harmonization of multiple reference models. In: 5th International Conference on Evaluation of Novel Approaches to Software Engineering, pp. 61–73. Springer, Berlin (2011)
13. Ashrafi, N.: The impact of software process improvement on quality: in theory and practice. Inf. Manag. **40**, 677–690 (2003)
14. Pardo, C., Pino, F.J., García, F., Piattini, M.: Harmonizing quality assurance processes and product characteristics. Computer **44**, 94–96 (2011)
15. García-Mireles, G.A., Moraga, M.Á., García, F., Piattini, M.: Towards the harmonization of process and product oriented software quality approaches. In: Winkler, D., O'Connor, R., Messnarz, R. (eds.) EuroSPI 2012. CCIS, vol. 301, pp. 133–144. Springer, Berlin (2012)
16. ISO: ISO/IEC FDIS 29110-2 Software engineering—lifecycle profiles for Very Small Entities (VSEs)—part 2: framework and taxonomy (2010)
17. ISO: ISO/IEC 12207 Systems and software engineering—software life cycle processes (2008)
18. Ferré, X., Juristo, N., Moreno, A.M.: Framework for integrating usability practices into the software process. In: 6th International Conference on Product Focused Software Process Improvement, pp. 202–215. Springer, Oulu (2005)
19. García-Mireles, G.A., Moraga, M.Á., Garcia, F., Piattini, M.: The influence of process quality on product usability: a systematic review. CLEI Electron. J. **16**, 1–13. http://www.clei.org/cleiej/paper.php?id=278 (2013)

Analysis of Coverage of Moprosoft Practices in Curricula Programs Related to Computer Science and Informatics

Mirna Muñoz, Adriana Peña Pérez Negrón, Jezreel Mejia and Graciela Lara Lopez

Abstract In Mexico, the small and medium size companies (SMEs) are key for the software development industry, in such a way that having highly qualified personal for the development of high quality software products is a fundamental piece to guarantee their permanency in the market. Therefore, matching the software industry requirements with the academy training represents a big problem that must be reduced for the benefit of both sectors. In this context, to understand the coverage of the academic curricula programs in higher education, regarding the software industry requirements is a fundamental aspect to take actions to reduce the current gap between industry and academy. This paper presents an analysis of coverage between Moprosoft norm, standard used for software industry to ensure quality in Software Engineering practices, and four academic curricula programs of higher education related to Computer Science and Informatics.

Keywords Moprosoft · Computer science and informatics curricula programs · Software industry · SMEs

M. Muñoz (✉) · J. Mejia
Centro de Investigación en Matemáticas, Avda. Universidad no. 222, 98068 Zacatecas, Mexico
e-mail: mirna.munoz@cimat.mx

J. Mejia
e-mail: jmejia@cimat.mx

A.P.P. Negrón · G.L. Lopez
Departamento de Ciencias de la Computación, CUCEI de la Universidad de Guadalajara, Av. Revolución No. 1500, 44430 Guadalajara, JAL, Mexico
e-mail: adriana.pena@cucei.udg.mx

G.L. Lopez
e-mail: graciela.lara@red.cucei.udg.mx

J. Mejia et al. (eds.), *Trends and Applications in Software Engineering*, Advances in Intelligent Systems and Computing 405,
DOI 10.1007/978-3-319-26285-7_4

1 Introduction

Software development in Small and Medium Enterprises (SMEs) has grown and strengthened, becoming a key element in the consolidation of the software industry [1–3]. According to the Mexican Association of Information Technology Industry [4], SMEs represent 87 % of the total of software development industries in Mexico. This fact highlights the importance of guaranteeing the quality of SMEs products.

In this context, providing qualified professionals able to work under quality models and standards represents a big challenge for universities. This challenge is not new, according to Moreno et al. [5] and Laporte [6, 7], one of the most critical tasks to be addressed in software education is to reduce the gap among educating software practitioners and the software development industry, to confront current and future challenges of the software industry.

Based on the research work made by Moreno et al. [5], the goal of this paper is to answer if the universities provide an adequate knowledge to Computer Science students to enable them to develop the skills needed to be integrated in organizations that work under process models.

In order to answer this question, this paper presents an analysis between the knowledge provided by the academy and the software industry requirements. For this, on the one hand we took four curricula programs of Mexican universities in Computer Science and Informatics such as: (1) BS in computer science; (2) software engineering; (3) degree in computer science; and (4) computer engineering were selected and analyzed; and on the other hand, the Moprosoft model, used by the software industry in Mexico, was selected and taken as a base of the software industry requirements.

After this introduction, the rest of the paper is structured as follows: Sect. 2 shows the background of this research work; Sect. 3 presents the followed methodology to analyze the curricula programs regarding the knowledge required to be able to perform the practices proposed by the Moprosoft model; Sect. 4 shows the obtained results; and finally, Sect. 5 presents conclusions.

2 Background

Two of the main challenges that graduate students in Computer Science and Informatics face when they are incorporated into an organizational environment are to be able: to perform a roll in a project development as part of a development team; and to work under quality models or standards, which are required by the organization, specifically in SMEs where the human resources are limited [8].

In this context, based on the work of Moreno et al. [5] and Laporte et al. [7], this research work aims to analyze if the knowledge provided at universities for computers science and informatics students is adequate, so that they can meet the requirements of the software industry regarding the quality models and standards.

In order to achieve this goal, this research began performing a comparative analysis, aimed to understand the coverage of the academic curricula programs in higher education based on what the government norms establish for their accreditation. Then, on the one hand, it was selected CONAIC [9], which is a Mexican government agency that seeks the quality assurance in educational programs of public and private institutions of education, specifically focusing on *the Framework for Accreditation of Academic Programs and Computing Higher Education*; and on the other hand, Moprosoft [10], which is a Mexican model developed to lead Mexican SMEs to improve the software development process and was established as the NMX-I059/02-NYCE-2011 norm.

The main results obtained of the analysis highlight the findings listed bellow. The complete analysis was published in [8].

(a) *Findings in requirement management*: the knowledge contained in the curricular model ANIEI allows having high coverage regarding the requirement management, but the knowledge regarding the requirement development has very low coverage. The knowledge of requirement development is important to identify in an adequate way the customer needs and to define them correctly.
(b) *Findings in project management*: the knowledge contained in the curricular model ANIEI allows having a good coverage regarding the project monitoring and control, but the coverage level regarding the practices of risk, validation, measurement and analysis, and configuration management should be improved. The knowledge related to risk management is very important, because it can affect the project; the knowledge related to validation is key to know if the developed product meets the customer needs. The knowledge related to measurement and analysis is necessary to choose the adequate measures to analyze the project performance. Finally, the knowledge related to the configuration management is necessary to control the base line of documentation and obtained products throughout the project development.

Based on the obtained results, as future work it was proposed to analyze specific curricula programs, such as: degree in computer science; software engineering; computer science; and computer engineering, as provided by Mexican universities: that is the goal of this paper.

3 Methodology for the Analysis of Curricula Programs

To perform the analysis of the four curricula programs of Mexican universities related to computer science and informatics such as: (1) degree in computer science; (2) software engineering; (3) computer science; and (4) computer engineering, it was taken as a base the methodology for analyzing coverage proposed in [8].

This methodology is composed of 3 phases: (a) analyze the Moprosoft model; (b) analyze the ANEI curricula and; (c) establish coverage. Because of this research goal, the methodology was adapted to meet it as follows:

(a) *Analyze Moprosoft model*: In this phase were selected the Moprosoft targeted processes, to identify the generic knowledge performed in each practice contained in the selected processes. This identified generic knowledge is taken for the analysis of the curricula programs.
(b) *Analyze the ANEI curricula*: In this phase were identified the ANEI curricular elements. This phase was completely changed because the analysis presented in this paper is focused on four curricula programs. Therefore, each subject of the total of each curricula program was analyzed in order to identify the knowledge provided.
(c) *Analyze coverage*: This phase was focused on establishing a scale of values as well as the coverage level between the CONAI and Moprosoft regarding the knowledge required to perform projects, which uses the Moprosoft norm. Then, the scale of values was used to analyze the four curricula programs evaluating the coverage level between the knowledge provided in the subject and the knowledge required to perform a Moprosoft practice. The scale of values was established in the range of 0–4 as follows:

- *0*: the knowledge provided through the subject does not have knowledge related to the Moprosoft practice. It means the practice has no coverage.
- *1*: the knowledge provided through the subject is minimal and indirectly related to the Moprosoft practice. It means that the practice has a low level of coverage.
- *2*: The knowledge provided through the subject is generic and useful to perform the Moprosoft practice. It means that the practice has a medium level of coverage.
- *3*: The knowledge provided through the subject directly supports the performance of the Moprosoft practice. It means that the practice has a high level of coverage.
- *4*: The knowledge provided through the subject is specific and directly related to the requirement to perform the Moprosoft practice. It means that the practice has a complete coverage level.

The same process proposed in [8] was followed in order to perform the analysis as shown in Fig. 1.

It is important to mention that the four authors execute the process individually, then, a crosschecking of the obtained results was performed in order to compare the values assigned by each researcher. Besides, a set of meetings was performed to get agreements in different values. In this way, the coverage values were refined. Next section shows the obtained results.

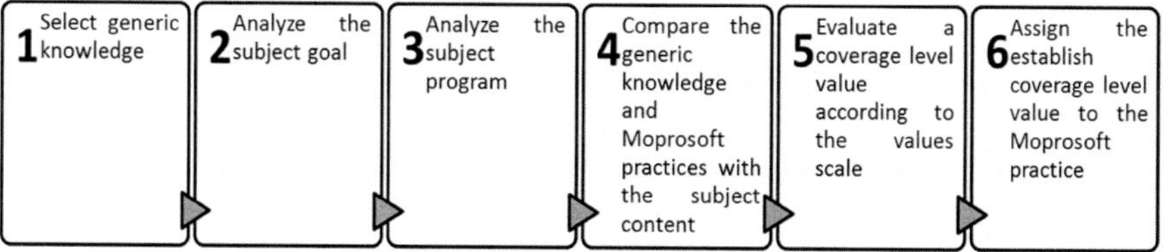

Fig. 1 Process followed to assign coverage level values to each Moprosoft practice

4 Obtained Results

It is important to mention that this study is focused in process related to project management, because it is a key process for organizations in order to achieve a maturity level 2 [11]. Processes related to project management are the most targeted in order to implement software improvements [11]. Also the processes that support the project management process included in the engineering and support areas are included.

Therefore, it is fundamental for undergraduate students, to be provided with the adequate knowledge to manage software processes covering the requirements to manage projects under the quality models and standards used in Mexican SMEs, such as the Moprosoft model.

The processes included in the analysis are: requirement management (REQM); project planning (PP); project monitoring and control (PMC); supplier agreement management (SAM); risk management (RSKM); requirements development (RD); verification (VER); validation (VAL); configuration management (CM); measurement and analysis (MA) among others.

The analyses performed are focused on two aspects:

- *Analysis of curricula program coverage by process*: to establish the coverage level by process. To achieve this, two steps were performed, first, individual coverage level for each subject of the curricula programs was calculated applying the next coverage value formula:

$$Coverage\,percentage\,value\,by\,practice\;=\Sigma i\ldots n\,practices\,coverage\,value/\\(number\,of\,practices\,*\,maximum\,coverage)$$

  where the maximum coverage level or complete coverage is 4.
- *Analysis comparing the four curricula program coverage by area*: to establish the coverage percentage by area. The process areas included are important for developing and supporting a project performance. To achieve it the next formula was applied:

$$Coverage\ percentage\ by\ area = \Sigma i\ldots n\ Coverage\ percentage\ value\ by\ practice/\ number\ of\ processes$$

Next, this section shows the analysis of the results obtained for each curricula program coverage by process (Sects. 4.1–4.5 shows an analysis comparing the four curricula programs coverage by area.

4.1 Analysis of BS in Computer Science Curricula Programs

The BS in computer science curricula program aims to create and maintain creative and innovative solutions regarding the information systems.

Table 1 shows the results obtained analyzing the BS in the computer science curricula programs. As can be seen in Table 1, risk management is the only process fully covered by the curricula program. The other process areas are not fully covered by the curricula program. However, it was found that graduates are better qualified to carry out the practices involved in requirement management, project monitoring and control, and requirements development.

Besides, process areas in which graduates needs to improve their training are: project planning, supplier management, verification, and validation.

4.2 Analysis of Software Engineering Curricula Program

The software engineering curricula program aims to train professionals in process development and the evolution of large and small scale software systems that solve problems in different areas, using appropriate tools to optimize time and cost.

Table 1 Summary of obtained values by process

Area	Processes	Percentage of coverage level (%)
Project management	Requirement management	92
	Project planning	76
	Project monitoring and control	80
	Supplier agreement management	38
	Risk management	100
Engineering	Requirements development	83
	Verification	59
	Validation	53
Support	Configuration management	68
	Measurement and analysis	50

Table 2 Summary of obtained values by process

Area	Processes	Percentage of coverage level (%)
Project management	Requirement management	100
	Project planning	84
	Project monitoring and control	96
	Supplier agreement management	88
	Risk management	100
Engineering	Requirements development	100
	Verification	89
	Validation	100
Support	Configuration management	85
	Measurement and analysis	100

Table 2 shows the results obtained of analyzing the software engineering curricula programs. As this table shows, processes such as requirement management, risk management, requirements development, validation, and measurement and analysis are processes fully covered by the curricula program.

The other process areas are not fully covered by the curricula program. However, it was found that graduates are better qualified to carry out the practices involved in project monitoring and control, verification, supplier agreement management, configuration management, and project planning.

4.3 *Analysis of Degree in Computer Science Curricula Program*

The degree in computer science curricula program aims to train professionals with analytical skills, critical skills, creativity and leadership to provide computational solutions in organizations applying information technology and communications.

Table 3 shows the results obtained of analyzing the degree in computer science curricula program. As this table shows, processes such as requirement management and requirements development are fully covered by the curricula program.

The other process areas are not fully covered by the curricula program. However, it was found that graduates are better qualified to carry out the practices involving validation, project monitoring and control, project planning, risk management, measurement and analysis, and verification.

Besides, process areas in which graduates need to improve their training are supplier agreement management and configuration management.

Table 3 Summary of obtained values by process

Area	Processes	Percentage of coverage level (%)
Project management	Requirement management	100
	Project planning	75
	Project monitoring and control	92
	Supplier agreement management	50
	Risk management	75
Engineering	Requirements development	100
	Verification	71
	Validation	94
Support	Configuration management	45
	Measurement and analysis	75

4.4 Analysis of Computer Engineering Curricula Program

The computer engineering curricula program aims to train professionals with analytic capacities, critical to provide creative solutions to the regional and state development using computer technology, and promoting socials values as well as the environmental care.

Table 4 shows the results obtained of analyzing the degree in computer science curricula program. As this table shows, not any process is fully covered by the curricula program. However, it was found that graduates are better qualified to carry out the practices involved in requirements development.

Process area in which graduates need to improve their training are: requirement management, verification, validation, and project planning; but they receive deficient training in processes such as project monitoring and control, risk management, and supplier agreement management.

Table 4 Summary of obtained values by process

Area	Processes	Percentage of coverage level (%)
Project management	Requirement management	50
	Project planning	30
	Project monitoring and control	26
	Supplier agreement management	13
	Risk management	25
Engineering	Requirements development	83
	Verification	54
	Validation	34
Support	Configuration management	33
	Measurement and analysis	55

4.5 *Analysis of the Four Curricula Programs Coverage Level by Area*

This analysis establishes the coverage level of curricula programs regarding three areas: project management, engineering and support.

Table 5 shows the obtained results comparing the four curricula programs as follows (column a) BS in computer science curricula program; (column b) software engineering curricula program; (column c) degree in computer science curricula program; and (column d) computer engineering curricula program.

The findings of the analysis performed to the obtained results are summarized as follows:

- *Project management*: it contains key processes to perform a project. This area analysis is focused on practices related to planning, monitoring, and controlling the project.
  Interpretation: not any area is fully covered by the curricula programs. However, it was found that graduates of software engineering curricula program (column b) are better qualified to carry out the practices involved in project management. However, it was found that graduates of BS in computer science curricula program (column a) and degree in computer science curricula program (column c) need to improve their training to carry out the practices involved in project management. Finally, graduates of computer engineering curricula program (column d) receive minimum training to carry out the practices involved in project management.
- *Engineering*: it contains processes that support the development of software. This area analysis is focused on practices related to verification, validation and requirement management.
  Interpretation: not any area is fully covered by the curricula program. However, it was found that graduates of software engineering curricula program (column b) and degree in computer science curricula program (column c) are better qualified to carry out the practices involved in engineering. However, it was found that graduates of BS in computer science curricula program (column a) and graduates of computer engineering curricula program (column d) need to improve their training to carry out the practices involved in engineering.
- *Support*: it contains processes focused on providing the necessary support to perform a project related to configuration management and, measurement and analysis.

Table 5 Summary of obtained percentage by area

Area	a (%)	b (%)	c (%)	d (%)
Project management	77	94	78	29
Engineering	65	96	88	57
Support	59	93	60	44

Table 6 Summary of obtained values by Moprosoft categories

Area	a (%)	b (%)	c (%)	d (%)
Management	67	83	76	22
Operation	70	90	77	32
Maintenance-operation	63	89	72	44

Interpretation: not any area is fully covered by the curricula program. However, it was found that graduates of software engineering curricula program (column b) are better qualified to carry out the practices involved in support. However, it was found that graduates of BS in computer science curricula program (column a), degree in computer science curricula program (column c) and graduates of computer engineering curricula program (column d) need to improve their training to carry out the practices involved in support.

Finally, Table 6 shows the obtained results comparing the four curricula programs of the coverage level of practices of Moprosoft categories: *management, operation and maintenance-operation* regardless of the area to which the practices belong, with a total of 37 practices of the management category; 41 practices of operation and 62 practices of maintenance-operation.

As Table 6 show, not any category is full covered by the four curricula program. However the four curricula programs have a tendency to train graduates in operation category. Besides, the graduates of software engineering curricula program (column b), degree in computer science curricula program (column c) and BS in computer science curricula program (column a) are better qualified to carry out the practices involved in management category; and graduates of computer engineering curricula program (column d) are better qualified to carry out the practices involved in maintenance-support categories.

5 Conclusions

This paper presents the results of analyzing the coverage of Moprosoft practices from the knowledge provided by four curricula programs of universities of México related to computer science and informatics such as: (1) BS in computer science curricula program; (2) software engineering curricula program; (3) degree in computer science curricula program; and (4) computer engineering curricula program.

The paper do not pretend to determine whether or not graduates are able to perform a practice, but to analyze how helpful is the knowledge provided in the four curricula programs for training graduates toward achieving the requirements of software industry regarding quality models and standards such as Moprosoft in México.

Therefore, it can be highlighted that software-engineering curricula program has a better coverage of practices involved in project management, engineering and support, followed by degree in computer science curricula program and BS in computer science curricula program, which have similar results, and finally computer engineering curricula program with the lowest results in the percentage of coverage level.

Besides, it is important to mention that the four curricula programs have a tendency to train graduates in the operation category.

Finally, it is important to mention that the obtained results are useful for both academic and industry sectors. For the academic viewpoint, it is highlighted the specific knowledge that should be improved in order to provide an adequate knowledge to undergraduates. And from the viewpoint of the software industry, they are highlighted the process areas in which graduates have received more or less training, useful for the design of training programs.

References

1. Moreno T.M.: Four Achilles Heel for SMEs, SME Observatory, Spanish papers online. http://www.observatoriopyme.org/index.php?option=com_content&view=article&id=74&Itemid=102 (2008)
2. Ministry of Industry: More information about the new definition of SMEs in EU, vol. 2012 (2012)
3. Ministry of Economy: SMEs: Fundamental Link to the Growth of Mexico, vol. 2013 (2013)
4. AMITI: Outline of Government Support for Software Industry, Mexican Association of Information Technology Industry (2010)
5. Moreno A., Sanchez-Segura M.A., Medina-Dominguez F.: Analysis of coverage of CMMI practices in software engineering curricula. In: SEPG Europe 2012 Conference proceedings, Special Report CMU/SEI-2012-SR-005, Sept 2012. pp. 42–77. (2012)
6. O'Connor, R.V., Mitasiunas, A. and Ross, M. (eds.) Proceedings of 1st International Workshop Software Process Education, Training and Professionalism (SPEPT 2015), EUR Workshop Proceedings Series, vol. 1368 (2015). http://ceur-ws.org/Vol-1368/
7. Laporte, C.Y., O'Connor, R.V., Garcia Paucar, L. and Gerancon B.: An innovative approach in developing standard professionals by involving software engineering students in implementing and improving international standards. J Soc. Stand. Prof. **67**(2), 2015. http://profs.etsmtl.ca/claporte/Publications/Publications/SES_2015.pdf
8. Mirna, M., Pérez-Negrón, A.P., Jezreel, M., Luis, C.-S.: Cobertura de la curricula unviersitaria para la industría de la Ingeniería del Software en México in the International conference of software process improvement (CIMPS 2014)
9. CONAIC "Marco de referencia para la Acreditación de programas académicos de Informática y Computación de Educación Superior" (2013) http://conaic.net/formatos/CRITERIOS_EVALUACION_CONAIC_2013.pdf
10. Octaba H. y Vázquez A.: MoProSoft: a software process model for small enterprises. In: Software Process Improvement for Small and Medium Enterprises, Techniques and cases studies, (Information Science Reference eds), p. 170 (2008)
11. Mirna, M., Gloria, G., Claudia, V.: Caracterizando las necesidades de las Pymes para Implementar Mejoras de Procesos Software: Una comparativa entre la teoría y la realidad, Revista Ibérica e Tecnologías de Información (RISTI), Número Especial No E1, pp. 1–15, (2014). ISSN 1646-9895

ProPAM/Static: A Static View of a Methodology for Process and Project Alignment

Paula Ventura Martins and Alberto Rodrigues da Silva

Abstract Process descriptions represent high-level plans and do not contain information necessary for concrete software projects. Processes that are unrelated to daily practices or hardly mapped to project practices, cause misalignments between processes and projects. We argue that software processes should emerge and evolve collaboratively within an organization. In this chapter we present a Process and Project Alignment Methodology for agile software process improvement and particularly describe its static view.

Keywords Software process improvement · Process management · Project management

1 Introduction

Software process improvement (SPI) is a challenge to organizations trying to continually improve software quality and productivity and to keep up their competitiveness [1]. Organizations tend to react to: (1) changes in the environment that they operate, (2) changes at a corporate level, (3) unplanned situations not considered in the model, or (4) improve the quality of their final products. Such changes may be caused, for example, by poor performance, by new tools acquired by the company to support its software development teams, changes in the marketing strategy or in clients' expectations and requirements. Thus, an existing process model must be modified or extended to reflect the evolution of the environment and/or internal changes. However, existing process models—that only take into account descriptive

P.V. Martins (✉)
Research Centre of Spatial and Organizational Dynamics, Universidade do Algarve, Campus de Gambelas, 8005-139 Faro, Portugal
e-mail: pventura@ualg.pt

A.R. da Silva
INESC-ID, Instituto Superior Técnico, Universidade de Lisboa, Lisbon, Portugal
e-mail: alberto.silva@tecnico.ulisboa.pt

J. Mejia et al. (eds.), *Trends and Applications in Software Engineering*, Advances in Intelligent Systems and Computing 405,
DOI 10.1007/978-3-319-26285-7_5

aspects, such as work related activities and technical work products—couldn't address such features. Additionally, several surveys and studies [2–4] have emphasized that the majority of small organisations are not adopting standards such as CMMI [5]. Another case is observed in Brazil where software industry and universities are working cooperatively in implementing a successful SPI strategy that take into account software engineering best practices and aligned to Brazilian software organizations context [6].

We argue that the emphasis in SPI should be stressed on communication, coordination, and collaboration within and among project teams in daily project activities, and consequently the effort in process improvement should be minimized and performed as natural as possible. Little attention had been paid to the effective implementation of SPI models, which has resulted in limited success for many SPI programs. SPI managers want guidance on how to implement SPI activities, rather than what SPI activities do actually implement. Limited research has been carried out on exploring new approaches to effectively implement SPI programs. However, to bridge this gap some initiatives have emerged, namely the MIGME-RRC methodology [7]. On this basis, we propose a new methodology to describe and improve software process based on organizations projects experience.

In this paper we propose a SPI methodology called Process and Project Alignment Methodology (ProPAM). The main goal is to develop a SPI methodology that can evolve with project's knowledge and consequent improvement at software development process level. This methodology takes into consideration project and organization's views and adapts the best practices to define a base software development process. The model is grounded in personal experience and observations from a real case study, related to the extensive literature review, and focused on the three main objectives: (1) further understand how modeling and implementation of software processes can contribute to successful SPI; (2) provide a SPI approach for SPI practitioners to ensure a successful process implementation and (3) contribute to the body of knowledge of SPI with a focus on implementation of software processes based on project experience.

The remainder of this paper is organized as follows: Sect. 2 presents related work and other initiatives. Section 3 overviews ProPAM and describes the details of its Static View. Finally, Sect. 4 presents the conclusions and discusses our perception that this proposal has innovative contributions for the community.

2 Related Work

There are several popular SPI models, such as ISO/IEC 15504 [8], CMMI [5] and ISO/IEC 29110 [9]. These models have become well known among practitioners and researchers. However, implementation and adoption of SPI at software development organizations is frequently unsuccessful [10, 11]. These SPI models are often prescriptive and attuned to those relative areas for which they are intended and therefore do not take into account other aspects like project and organization specific features. Rather than just repair and adjust the process to specific areas imposed by these SPI models, we should refocus SPI to analyze the current organization practices and introduce practices adapted to the organizations needs. SPI should be based on project's experience and learn with project team member's. Software development is not a rigid or a disciplined manufacture. It has a strong creative and social interaction that can't be totally re-planned in a standardized and detailed process model elaborated by specific groups and without active participation of project teams. These SPI models identify **what** to improve but don't give any information about **how** to do it. Indeed, given the reported problems with existing SPI models, there is a need for a more comprehensive model to SPI.

We argue that SPI requires further researches on SPI models based on project's experience. Following the trend of agile processes [12, 13], SPI requires a commitment to project learning and this organizational knowledge is constructed through collaboration of project team members. Therefore, we need to include guidelines in a SPI model that allow to incorporate project team knowledge in the software process (without constraints imposed by a standard which limits embed tacit knowledge) and that can address features not focused on existing SPI models.

3 Process and Project Alignment Methodology—The Static View

This section presents an overview and details of the ProPAM methodology [15]. ProPAM is about how the process and project are represented and how project teams acquire and use knowledge to improve work. ProPAM is different from existing models in which SPI is seen as starting with the implementation of best practices according to a predetermined scheme, independently of organization's experience of problematic situations. The proposed methodology proposes solving the problems faced in software development projects carried within the organizations. A critical feature in ProPAM is the integration of SPI activities with software development activities. This way, we considered project teams and projects as the baseline for improvement. Project managers and project teams, under the supervision of the process manager, are the foremost responsible for keeping the organization's processes on the leading edge. As Fig. 1 illustrates, ProPAM includes SPI activities to monitoring and tracking software projects (project level) besides the SPI activities that intend to develop and implement the software process (process level). The main differences between these two levels are the followings.

At **project level**, ProPAM helps organizations in their efforts to assess and manage problematic situations of specific projects, and to develop and implement solutions to manage these problems. The project level encompasses project(s) information needed to systematically support or reject many of the decisions about the process. At project level, team members work together to develop work products. This focus on project team and their collaborative process is important because no one embodies the breadth and depth of knowledge necessary to

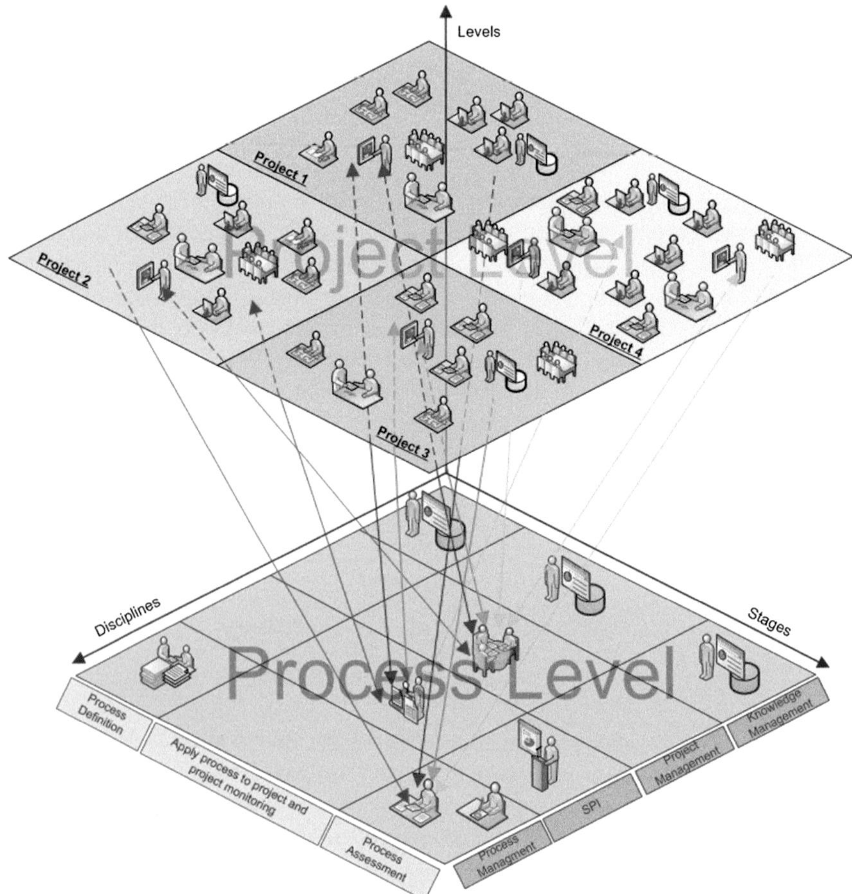

Fig. 1 ProPAM levels

comprehend large and complex software systems. Project teams are concerned with concrete situations as experienced in all their complexity during software development. Projects context is constantly being created and recreated and it can't be based on a static process model. Participating in a project team is consequently not only a matter of developing software, but also to change organization's knowledge about software development.

At **process level**, project's feedbacks conduct to process reviews and iterative process improvement. The dynamic interplay between these two levels shows the synergy between the activities performed by project roles (project manager and team member) and the activities performed by the process roles (process manager) involved in SPI. At process level, actors involved in SPI programs take time to express its shared practices in a form that can meaningfully be understood and exploited by other organizational actors. This includes not only the definition of concepts, models and guidelines, but also the evaluation of success of the improvements.

The scope of the levels is defined considering that process and projects actors collaborate on SPI programs. However, to manage the inherent complexity of these levels, namely ProPAM represented at process level, it is common practice to present views on the models of a level. In general, a view is defined as a projection of a process model that focuses on selected features of the process [14]. ProPAM is organized in two correlated and complementary views, the static view and the dynamic view that represent the behavior at that particular level. Whereas the static view describes aspects of the methodology as core and supporting disciplines in terms of activities, work products and roles, the dynamic view shows the lifecycle aspects of ProPAM expressed in terms of stages and milestones. The remaining of the section is dedicated to details of the static view. Previous work already described the dynamic view [16].

The ProPAM static view describes disciplines involved in SPI and relations between them. Static view is expressed as workflow diagrams, which show structural elements (roles, work products and activities) involved in each ProPAM discipline. Swim lanes in the workflow diagram make obvious the roles responsible to perform specific activities and also identifies involved input and output work products. For each role, control flow transitions between activities are omitted since activities are neither performed in sequence, nor done all at once. Nevertheless, such representation does not describe SPI program changes with time passing. A time-based perspective of the process is left to dynamic view.

ProPAM static view integrates project management, process management, SPI and Knowledge Management (KM) disciplines. These disciplines assure alignment of projects with organization vision and goals, and the adopted and improved software process. Other disciplines of concern were omitted, like business modeling, analyze and design, environment, requirements management or configuration management, because those concerns are considered too specific for SPI programs.

Project Management. Project managers are usually interested to be informed about how the project follows its base process and how to handle the changes introduced in the project that are not compliant with the respective process. It is important to detect deviations from schedules (project control and project tracking activities) as soon as possible in order to take corrective actions. Deviations allow identifying elements that do not appear or are incorrectly described in the software process. Therefore, project managers have to be informed about process states in a way that satisfies management needs. This bridges the gap between process management and project management, since project plans should reflect the exact set of activities defined for a given process. To avoid creating detailed plans, project managers may create the plan incrementally, and using only higher-level activities, leveraging lower level tasks only as a guide for how to do the work. The most important goal is to address conflicts and align projects and processes. Figure 2 illustrates the main roles, activities and work products involved in the Project Management discipline.

Software Process Management. Software process management discipline involves actions performed to coordinate knowledge acquisition about software

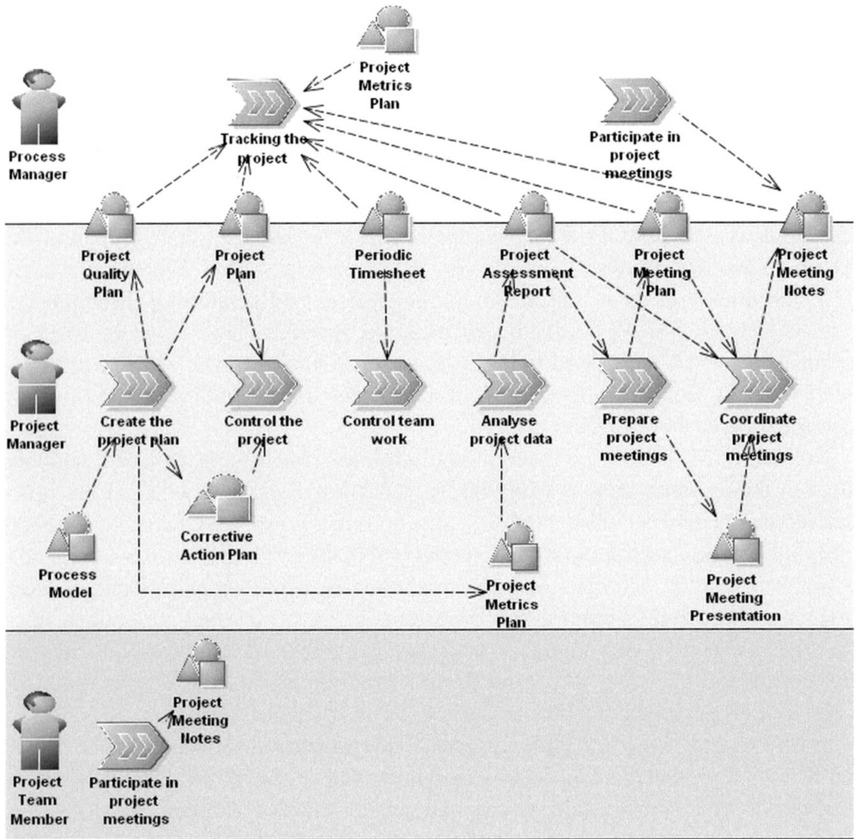

Fig. 2 Project management view

processes, to model and to analyze the way teams develop software and, finally, to ensure that future software processes are carried out on the basis of findings obtained in process analysis [17]. Software process management is a collective work involving project managers, senior engineers and the process manager. Nevertheless, at process level the process manager must be concentrated in process definition and implementation. While at project level the process manager coordinate the interaction with projects team members with respect to process assessment. Software process roles should develop the following activities with direct impact on SPI: (1) collect relevant material; (2) organize interviews and questionnaires; (3) make interviews; (4) understand project experiences; (5) define and implements the process model; (6) establish engineering practices; (7) identify the technical infrastructure; and (8) participate in interviews/answer to questionnaires. Figure 3 presents the main roles, activities and work products involved in the Software Process Management discipline. Some details of the main activities allow understanding the importance of this discipline. In this case, the project manager has the same responsibilities of the other team members, so he isn't seen as a specific role.

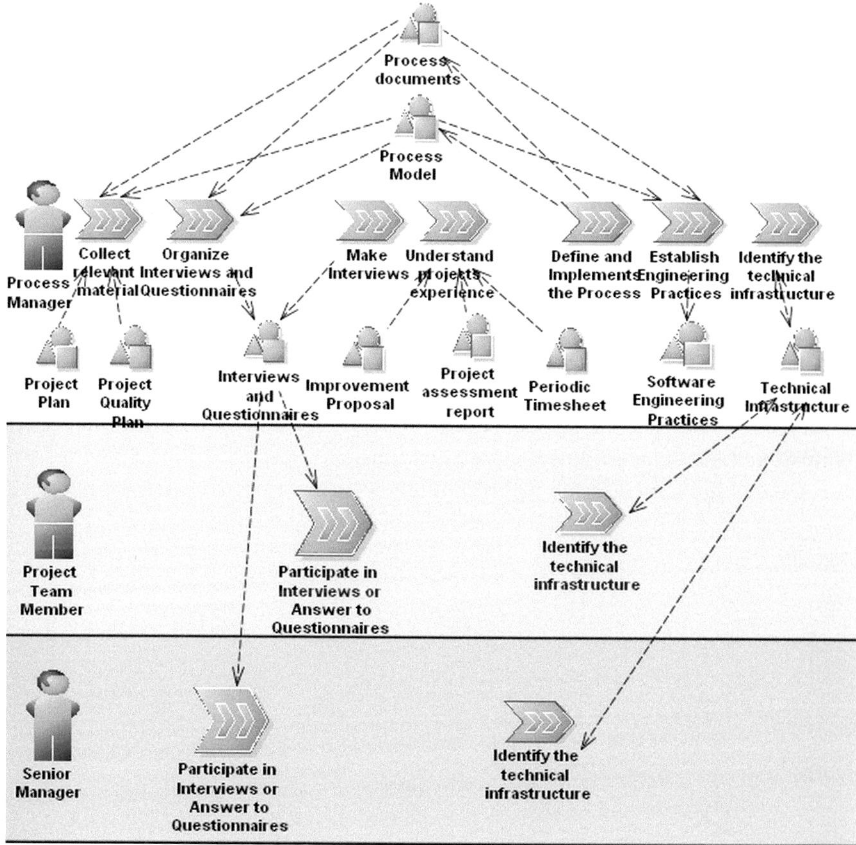

Fig. 3 Software process management view

The most important goal is to design a set of solutions for the software process based on performed projects. To help the viewer understanding the diagram in Fig. 3, a restriction some flows from and to work products were omitted, since these work products are inputs or outputs of almost all the activities of the discipline.

Software Process Improvement. The effort of supporting software processes is encompassed by the SPI discipline of the ProPAM methodology. This discipline extends the process management discipline, where the main difference is the scope: the process management discipline is concerned with the process configuration for the organization, while the SPI discipline addresses improvements in the process itself based on assessment results. SPI is the discipline of characterizing, defining, measuring and improving software management and development processes, leading to software business success, and successful software development management. Success is defined in terms of greater design innovation, faster cycle times, lower development costs, and higher product quality, simultaneously [18].

SPI focus is related to the establishment of a set of responsible roles and of the associated competences concerned to the software development process with the aim of improving the organization's software process. The main activity of this discipline is the maintenance of software process knowledge and the improvement of coordination and monitoring activities. The organization must plan to create a stable environment and monitor these activities in order to have clear commitments for current and future projects. The most important goals to be achieved are: (1) software development process and improvement activities are coordinated throughout the organization; (2) the strengths and weakness of the used software process are identified relative to a base process, if it was previously defined; and (3) improvement activities are always planned. ProPAM also suggests that organizations should identify a group of software managers composed by skilled persons and (internal or external to the organization) advisors, who contribute to identify the process strengths and to improve it when weakness are identified. Figure 4 presents the workflow diagram that illustrates the main roles, activities and work products involved in the SPI discipline.

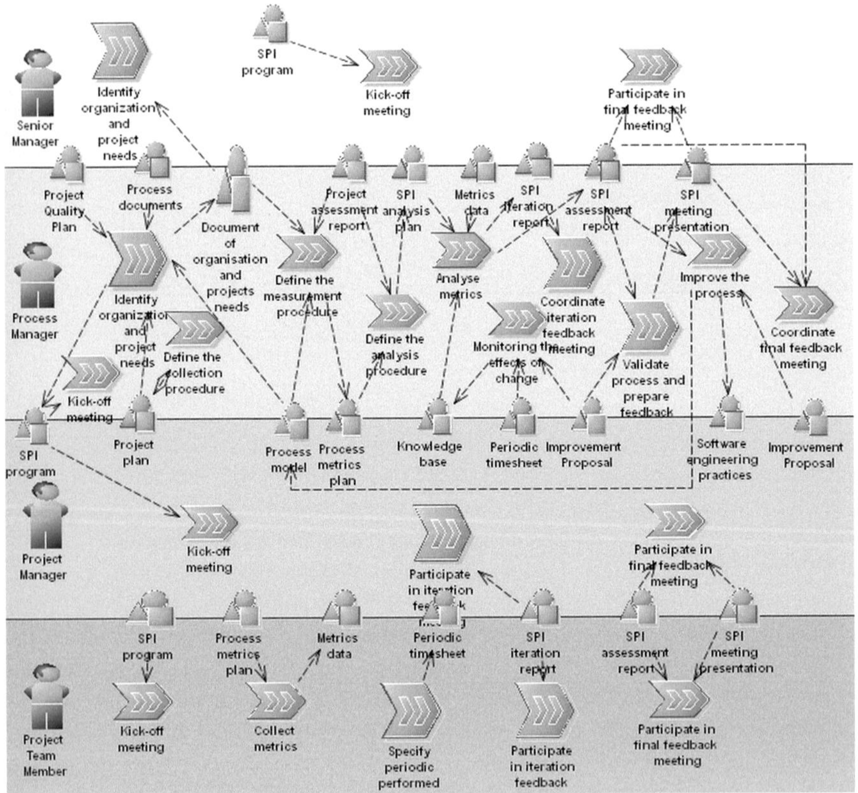

Fig. 4 Software process improvement view

Knowledge Management (KM). Data is organized into information by combining data with prior knowledge and the person's self-system to create a knowledge representation. This is normally done to solve a problem or make sense of a phenomenon. This knowledge representation is consistently changing as we receive new inputs, such as learning, feelings, and experiences. Knowledge is dynamic, that is, our various knowledge representations change and grow with each new experience and learning. Due to the complexity of knowledge representations, most are not captured by documents; rather they only reside within the creator of the representation. In many cases, the knowledge representation stays within the creator, in which case the "flow of knowledge" stops.

A KM system, which may be as simple as a story or as complex as an expensive computer program, captures a snapshot of the person's knowledge representation. Others may make use of the knowledge representation "snapshot" by using the story or tapping into the KM system and then combining it with their prior knowledge. This in turn forms a new or modified knowledge representation. This knowledge representation is then applied to solve a personal or business need, or explain a phenomenon. The main goal is to connect knowledge providers with seekers concerning software processes. Figure 5 presents the main roles, activities and work products involved in the Knowledge Management discipline.

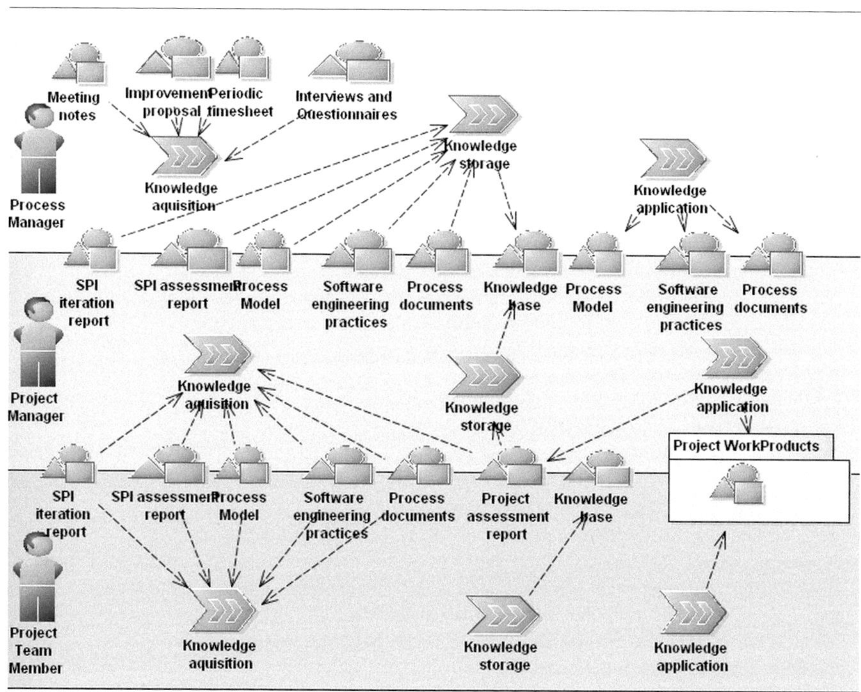

Fig. 5 Knowledge management view

4 Conclusion

In the state of the art, several problems were identified concerning SPI programs based on standard SPI models. ProPAM is presents as an alternative approach to SPI focused on gaps and problems identified on existing SPI standards.

A case study was performed in a small organization without conditions to accomplish a maturity assessment using CMMI. The main goal of the case study was to give us insight into how SPI initiatives can best suit organizational goals and also showed us the impact on the organization and the strengths and weaknesses of ProPAM. Since the focus of the paper is to present ProPAM as an alternative agile SPI approach, the description of the case study is out-of-scope.

As final conclusion, the prescriptive nature of traditional approaches (such as CMMI) and costs necessary to implement SPI programs are the main reasons for further research on SPI based on project's experience. Namely, SPI models must address the importance of using the experience of software teams as an important source to SPI. Another gap observed was the deficient alignment between the process and projects. Nevertheless, the contribution of this work wasn't just an approach to align process and project specifications; we also proposed a mechanism to analyze evolution based on the changing needs of the software development organization.

Acknowledgments This work was partially supported by national funds through FCT (Foundation for Science and Technology) through projects UID/SOC/04020/2013 and UID/CEC/50021/2013.

References

1. Salo, O.: Improving software development practices in an agile fashion. In: Book Improving Software Development Practices in an Agile Fashion, Series Improving Software Development Practices in an Agile Fashion, Agile-ITEA, p. 8 (2005)
2. Staples, M., et al.: An exploratory study of why organizations do not adopt CMMI. J. Syst. Softw. **80**(6), 883–895 (2007)
3. Coleman, G., O'Connor, R.V.: Investigating software process in practice: a grounded theory perspective. J. Syst. Softw. **81**(5), 772–784 (2008)
4. Basri, S., O'Connor, R.: Organizational commitment towards software process improvement an Irish software VSEs case study. In: Proceedings 4th International Symposium on Information Technology 2010 (ITSim 2010) (2010)
5. SEI, CMMI for Development, Version 1.3, S. E. I.-C. M. University (2010)
6. Weber, K. et al.: MPS model-based software acquisition process improvement in Brazil. In: Proceedings 6th Quality of Information and Communications Technology (QUATIC 2007), pp. 110–122. IEEE Computer Society, Lisbon (2007)
7. Mirna, M. et al.: The results analysis of using MIGME-RRC methodology for software process improvement. In: Proceedings 6th Iberian Conference on Information Systems and Technologies (CISTI) (2011)
8. ISO/IEC, 15504-2 Information technology—software process assessment—part 2: a reference model for processes and process capability, ISO/IEC TR 15504-2, July, 1998

9. Laporte, C.Y. et al.: A software engineering lifecycle standard for very small enterprises. In: Proceedings Software Process Improvement, 15th European Conference, EuroSPI 2008, pp. 129–141. Springer, Berlin (2008)
10. Goldenson, D., Herbsleb, J.D.: After the appraisal: a systematic survey of process improvement, its benefits, and factors that influence success (1995)
11. Baddoo, N., Hall, T.: De-motivators for software process improvement: an analysis of practitioners' views. J. Syst. Softw. **66**(1), 23–33 (2003)
12. Beck, K.: Extreme programming explained: embrace change, p. 224. Addison Wesley, Boston (2004)
13. Rising, L., Janoff, N.: The Scrum software development process for small teams. IEEE Softw. **17**, 26–32 (2000)
14. Verlage, M.: Multi–view modeling of software processes. In: Proceedings of the Third European Workshop on Software Process Technology, pp. 123–126. Springer, Berlin (1994)
15. Martins, P.V., Silva, A.R.: ProPAM: discussion for a new SPI approach. Softw. Qual. Prof. J. **11**(2), 4–17 (2009) (American Society for Quality)
16. Martins, P.V., Silva, A.R.: ProjectIT-Enterprise: a software process improvement framework. In: Industrial Proceedings of the 17th EuroSPI Conference, pp. 257–266. Grenoble, France (2010)
17. Dissmann, S. et al.: Integration of software process management and development history recording. In: Proceedings Second Asia-Pacific Software Engineering Conference (APSEC'95), p. 468 (1995)
18. Rico, D.: Using cost benefit analyses to develop software process improvement (SPI) strategies, A. E. S. D. ITT Industries, Defense Technical Information Center (DTIC)/ AI (2000)

Proposal of a Hybrid Process to Manage Vulnerabilities in Web Applications

Ana L. Hernández-Saucedo and Jezreel Mejía

Abstract The information systems security is essential for organizations because organizations use information systems to manage their key information related to customers, products, and transactions, among others. The information systems of organizations are mostly web. However, over 70 % of the vulnerabilities are found in web applications, such as SQL Injection, Cross-site Scripting (XSS), Cross Site Request Forgery CSRF, Insecure Configuration Management, among others. Therefore, it is very important to secure the web systems. Therefore in the last 3 years have been observed an increase in the vulnerabilities having impact in the web systems attacks. Moreover, it has been detected that organizations do not implement procedures or processes to manage vulnerabilities, leaving exposed their systems. In this context, this paper presents a hybrid process that will enable organizations to detect and manage vulnerabilities in their web applications.

Keywords Security of information systems · Vulnerabilities · Models and standards · Web applications

1 Introduction

Today the information system security is essential for all organizations, because they manage key information related to customer and clients, services, among others [1, 2]. Most organizations use web applications in order to publicize their information,

A.L. Hernández-Saucedo (✉) · J. Mejía
Centro de Investigación en Matemáticas, Av. Universidad No. 222, 98068 Guadalupe, Zacatecas, Mexico
e-mail: ana.hernandez@cimat.mx

J. Mejía
e-mail: jmejia@cimat.mx

J. Mejia et al. (eds.), *Trends and Applications in Software Engineering*,
Advances in Intelligent Systems and Computing 405,
DOI 10.1007/978-3-319-26285-7_6

services and products [1], however more than 70 % of the vulnerabilities are detected in the application layer [3, 4] and 75 % of attacks are focused on applications [3].

In this context, statistics in [5, 6] indicate that the vulnerabilities have increased in the last three years. In 2011 were recorded 4000 vulnerabilities reports compared with over 5000 vulnerabilities in 2014. Besides, the most common vulnerabilities in web applications are included: place Cross-Site scripting XSS vulnerabilities, Denial of Service DoS, SQL Injection, Cross-Site Request Forgery (CSRF) and finally file inclusion vulnerabilities [7].

The increase in the vulnerabilities has an impact in the increase of attacks because use of new employed attack techniques [8]. Therefore, the systems must be free of vulnerabilities. Then, it is important to detect and mange vulnerabilities to prevent and avoid the stealing or alteration of any information included in these systems.

However, organizations do not perform or implement procedures or processes to detect and manage vulnerabilities of their web systems, as well as, don't implement good security practices such as the practices proposed by the CERT-RMM model or ISO/IEC 27002 standard, leaving their systems exposed to receive a potential attacks.

As a solution, this paper presents a proposal of a hybrid process that will enable organization to detect and manage vulnerabilities of their web applications. To achieve this, the next steps were performed: (1) Conduct a systematic review to identify the current vulnerabilities, the main tools and techniques that allow detecting vulnerabilities and the main models or standards focuses on information security. (2) Perform a traceability to define the hybrid process from models and standards that allow managing vulnerabilities.

This paper is structured as follow; Sect. 2 presents the main results of the systematic review as a basis for establishing the hybrid process, then models and standard that allow vulnerability management. Section 3 presents the steps performed to traceability in order to establish the proposed hybrid process for managing vulnerabilities in web applications, finally, Sect. 4 presents the conclusions.

2 Background

The systematic literature review protocol was conducted to find the current vulnerabilities, the main tools and techniques that allow detecting vulnerabilities and identify the main models or standards focused on information security.

2.1 Systematic Review

A systematic literature review was conducted at the end of 2014 following the guidelines to perform systematic literature reviews proposed by Kitchenham [9]. As a result of the systematic review the main techniques and tools used to detect vulnerabilities in web applications were obtained. The protocol and steps of the systematic review were published in the journal Recibe at the beginning of 2015 [10]. Table 1 shows a part of main systematic review's results regarding the main techniques and tools [10].

2.2 *Models and Standards*

As another result of the systematic review, we found that the CERT-RMM model and ISO 27002 standard are focused on security, specifically in vulnerability management. These models and standards provide a guide with activities to be

Table 1 Traceability of attack, vulnerability, techniques and tools (systematic review results)

Attack	Vulnerability	Technique vulnerabilities detection	Tool for vulnerabilities detection
SQL injection	Injection	Static code analysis	QualysGuard Web Application Scanning WAS
		Dynamic code analysis	WebSite Security Audit- WSSA
		Penetration testing	• WEBAPP 360: Enterprise class web application scanning
			• Retina web security scanner
			• W3AF Web Application Attack and Audit Framework
Session fixation attack	Broken authentication and session management	Session management standard	• QualysGuard Web Application Scanning WAS
			• WebSite Security Audit-WSSA
			• Retina web security scanner
			• WEBAPP 360: enterprise class web application scanning
			• W3AF Web Application Attack and Audit Framework
XSS attack	Cross-site scripting (XSS)	Static code analysis	• QualysGuard Web Application Scanning WAS
		Penetration testing	• WebSite Security Audit-WSSA
			• Retina web security scanner
			• WEBAPP 360: enterprise class web application scanning
			• W3AF Web Application Attack and Audit Framework
			• OWASP Xenotix framework

performed to manage vulnerabilities in a properly way, allow to achieve an optimal level of security for the organization [11]. The next subsection the CERT-RMM and ISO 27002 are briefly described.

2.2.1 CERT-RMM

CERT Resilience Management Model (CERT-RMM) aims to manage the operational recovery resilience at a level that supports the mission success. CERT-RMM is not based on the CMMI model; also it does not form an additional CMMI constellation or directly intersect with existing constellations. However, CERT-RMM uses several CMMI components, including fundamental process areas and process areas from CMMI-DEV [12]. CERT-RMM has 26 process areas that are organized into categories of high level of operational resilience: engineering, enterprise management, operations and process management. In this model, the process area focused on vulnerability management is the vulnerability analysis and resolution that is part of the operations category. The purpose of vulnerability analysis and resolution process area is to identify, analyze, and manage vulnerabilities in an organization's operating environment. Vulnerability analysis and resolution informs the organization of threats that must be analyzed in the risk management process to determine whether they represent a tangible risk to the organization based on its unique risk drivers, appetite, and tolerance [12]. The Table 2 shows the practices in this process area.

2.2.2 ISO/IEC 27000

International standard for management systems provides a model for the implementation and operation of a management system. With the use of system management for information security included in this standard, organizations can develop and implement a model to manage the security of their assets and prepare an independent evaluation [13]. This standard provides an overview of the management system information security, as well as the terms and definitions commonly used in the SGSI family of standards.

As part of the SGSI family of standards ISO/IEC 27001 and ISO/IEC 27002 among others are included. The standard ISO/IEC 27002 establishes guidelines and general principles for beginning, implementation, maintenance and improvement the information security management [14]. Their goal is to provide information for implementing security information in an organization. It is a set of best practices for developing and maintaining safety standards and management practices in an organization to improve the information security reliability [15, 16].

In this standard, the domain focuses on vulnerability management is the operations security domain. This domain contains control objective of technical vulnerability management, its objective is to prevent the exploitation of technical vulnerabilities. The next section shows the implementation guide of this control objective.

Table 2 Activities in the vulnerability analysis and resolution process area [12]

Specific goals	Practices	Subpractices
VAR:SG1 prepare for vulnerability analysis and resolution	VAR:SG1.SP1 establish scope	• Identify the assets that are the focus of vulnerability analysis and resolution activities
		• Identify the operational environments where vulnerabilities may exist for each asset
	VAR:SG1.SP2 establish a vulnerability analysis and resolution strategy	• Define the scope of vulnerability analysis and resolution activities
		• Develop and document an operational vulnerability analysis and resolution strategy
		• Communicate the operational vulnerability analysis and resolution strategy to relevant stakeholders and obtain their commitment to the activities described in the strategy
		• Assign resources to specific vulnerability analysis and resolution roles and responsibilities
		• Identify the tools, techniques, and methods that the organization will use to identify vulnerabilities to assets
VAR:SG2 identify and analyze vulnerabilities	VAR:SG2.SP1 identify sources of vulnerability information	• Identify sources of relevant vulnerability information
		• Review sources on a regular basis and update as necessary
	VAR:SG2.SP2 discover vulnerabilities	• Discover vulnerabilities
		• Provide training to the staff to perform data collection and discover vulnerabilities
		• Populate the vulnerability repository
		• Provide access to the vulnerability repository to appropriate process stakeholders
	VAR:SG2.SP3 analyze vulnerabilities	• Develop prioritization guidelines for vulnerabilities
		• Analyze the structure and action of the vulnerability
		• Prioritize and categorize vulnerabilities for disposition
		• Update the vulnerability repository with analysis and prioritization and categorization information

(continued)

Table 2 (continued)

Specific goals	Practices	Subpractices
VAR:SG3 manage exposure to vulnerabilities	VAR:SG3.SP1 manage exposure to vulnerabilities	• Develop a vulnerability management strategy for all vulnerabilities that require resolution
		• Ensure that relevant stakeholders are informed of resolution activities
		• Update the vulnerability repository with information about the vulnerability management strategy
		• Monitor the status of open vulnerabilities
		• Analyze the effectiveness of vulnerability management strategies to ensure that objectives are achieved
VAR:SG4 identify root causes	VAR:SG4.SP1 perform root-cause analysis	• Identify and select root-cause tools, techniques, and methods appropriate to be used in analyzing the underlying causes of vulnerabilities
		• Identify and analyze the root causes of vulnerabilities
		• Develop and implement strategies to address root causes
		• Monitor the effects of implementing strategies to address root causes

Implementation Guide to Establish Control Objectives in Technical Vulnerabilities

The control objective of technical vulnerability management has the controls of Management of technical vulnerabilities and restriction on software installation. The activities to the implementation guideline are:

- A current and complete inventory of assets is a prerequisite for effective technical vulnerability management.
- A current and complete inventory of assets is a prerequisite for effective technical vulnerability management.
- Define and establish the roles and responsibilities associated with technical vulnerability management, including vulnerability monitoring, vulnerability risk assessment, patching, asset tracking and any coordination responsibilities required.
- Information resources that will be used to identify relevant technical vulnerabilities and to maintain awareness about them should be identified for software and other technology.
- Defined to react to notifications of potentially relevant technical vulnerabilities

- Once a potential technical vulnerability has been identified, the organization should identify the associated risks and the actions to be taken; such action could involve patching of vulnerable systems or applying other controls.
- Depending on how urgently a technical vulnerability needs to be addressed, the action taken should be carried out according to the controls related to change management or by following information security incident response procedures
- If a patch is available from a legitimate source, the risks associated with installing the patch should be assessed
- Patches should be tested and evaluated before they are installed to ensure they are effective and do not result in side effects that cannot be tolerated; if no patch is available, other controls should be considered, such as:
  - Turning off services or capabilities related to the vulnerability
  - Adapting or adding access controls, e.g. firewalls, at network borders
  - Increased monitoring to detect actual attacks
  - Raising awareness of the vulnerability
- An audit log should be kept for all procedures undertaken
- The technical vulnerability management process should be regularly monitored and evaluated in order to ensure its effectiveness and efficiency
- Systems at high risk should be addressed first
- An effective technical vulnerability management process should be aligned with incident management activities, to communicate data on vulnerabilities to the incident response function and provide technical procedures to be carried out should an incident occur
- Define a procedure to address the situation where a vulnerability has been identified but there is no suitable countermeasure

The implementation guideline restriction activities in the installation of software control are:

- Define and enforce strict policy on which types of software users may install
- The principle of least privilege should be applied. If granted certain privileges, users may have the ability to install software.

3 Vulnerability Management Process in Web Applications

According to the results of the systematic review a hybrid process is proposed using the CERT-RMM model and the ISO/IEC 27002 standard for managing vulnerabilities in web applications. A hybrid is all that is the result of mixing two or more elements of different types. A hybrid borrows from all component elements into something new [17].

The hybrid vulnerability management process was established by one activity. This activity is related to the definition of hybrid process between CERT-RMM model and ISO/IEC 27002 standard. This activity includes the following steps:

3.1 Steps to Develop Hybrid Process

In order to establish the proposed hybrid process the following steps between CERT-RMM model and ISO/IEC 27002 standard were performed:

1. Analyze of the official information of CERT-RMM model
   a. Input artifact: Official documentation of CERT-RMM model
2. Analyze of the official information of ISO/IEC 27002 standard.
   a. Input artifact: Official documentation of ISO/IEC 27002 standard
3. Identify process areas and domains related with vulnerability management
   a. Output artifact: List of process areas and domains related with vulnerability management
4. Establish of Traceability: Identification of the similarities and differences of both structures. It focused the analysis on process areas and domains
   a. Output artifact: List of similar activities that are different
   b. Structure analyze: process structure (practice or activities, process area, specific goals, practices and subpractices)
5. Analyze whether the activities that are different are not implicit in other activities
6. Define of the hybrid process using activities of CERT-RMM model and ISO/IEC 27002 to complement the activities of vulnerability management according to our needs.
   a. Output artifact: Documentation of hybrid process

According to the activities listed above, Table 3 shows part of obtained traceability. This traceability was performed among the vulnerability analysis and resolution of CERT-RMM process area that are shown in Table 2 and the implementation guidance of control objective technical vulnerability management of ISO/IEC standard that are shown in Sect. 2.2.2.1.

As a result of the performed traceability the hybrid process was defined as follows:

1. Identify and records the assets to be analyze to detect vulnerabilities. Record information that supports vulnerability management, such as version number, current state of implementation, responsible person, and others.
   a. Output artifact: Document the organization's assets, as well as information that support vulnerability management.
2. Define and assign the responsibilities for vulnerability management.
   a. Output artifact: Document of responsibilities definition and assignment

Table 3 Traceability of CERT-RMM and ISO/IEC 27002 standard

Specific goals	Practices	Subpractices	ISO/IEC 27002
VAR:SG1 prepare for vulnerability analysis and resolution	VAR:SG1.SP1 establish scope	• Identify the assets that are the focus of vulnerability analysis and resolution activities	A current and complete inventory of assets is a prerequisite for effective technical vulnerability management
		• Identify the operational environments where vulnerabilities may exist for each asset	
		• Define the scope of vulnerability analysis and resolution activities	
	VAR:SG1.SP2 establish a vulnerability analysis and resolution strategy	• Develop and document an operational vulnerability analysis and resolution strategy	Define and establish the roles and responsibilities associated with technical vulnerability management, including vulnerability monitoring, vulnerability risk assessment, patching, asset tracking and any coordination responsibilities required
		• Communicate the operational vulnerability analysis and resolution strategy to relevant stakeholders and obtain their commitment to the activities described in the strategy	
		• Assign resources to specific vulnerability analysis and resolution roles and responsibilities	
		• Identify the tools, techniques, and methods that the organization will use to identify vulnerabilities to assets	Information resources that will be used to identify relevant technical vulnerabilities and to maintain awareness about them should be identified for software and other technology

3. Identify tools and techniques to be used to identify vulnerabilities.
    a. Output artifact: List of tools and techniques for identifying vulnerabilities
4. Identify resources or sources information to be used to identify vulnerabilities
    a. Output artifact: List of information source to identify vulnerabilities
5. Develop guidelines to prioritize the vulnerabilities. Vulnerabilities with a higher risk should be treated first.
    a. Output artifact: Guide for prioritizing vulnerabilities

6. Discover vulnerabilities. Update the vulnerabilities repository according to the discovered vulnerabilities
7. Identify the risks associated with each discovered vulnerability
   a. Output artifact: List of vulnerabilities risk
8. Develop actions to address vulnerabilities that require solution. If no actions are available to resolve the vulnerability, consider:
   a. Disable the capabilities related to the vulnerability
   b. Increase monitoring to detect or prevent attacks
   c. Increase awareness of the vulnerability
   d. Monitor information source for possible actions to perform
9. Evaluate the risks associated with the actions taken to address vulnerabilities.
10. Update the information vulnerabilities repository with actions or procedures performed to address the vulnerability.
11. Monitor the status of the vulnerability
12. Identify and analyze the root-causes of vulnerability
    a. Output artifact: Document the root-causes of vulnerability
13. Develop and implement strategies to address the root-causes.
    a. Output artifact: Document of strategies to address the root-cause of vulnerability
14. Monitor effectiveness of the implemented strategies to address the root-causes.

The proposed hybrid process takes into account activities to reduce the risk resulting from exposure of published technical vulnerabilities and allow identify, analyze and manage the vulnerabilities in an organization's operations environment.

4 Conclusions

The information system security is essential to ensure the integrity of information in organizations. A great amount of this information is in their web systems. This research shows that a high proportion of the vulnerabilities are in the applications layer. For this reason it is very important to perform vulnerability detection in addition to establish processes or procedures for managing these vulnerabilities.

As result of the systematic review it was observed that there are researches that disclose techniques and tools to detect vulnerabilities; however it was appreciated through the analysis of these results that there a lack of knowledge about models and standards to manage these vulnerabilities.

Besides, it was found a model and a standard that provide guidance for managing vulnerabilities. The CERT-RMM model focuses on operational vulnerabilities and ISO/IEC 27002 standard focuses on managing technical vulnerabilities. To cover both kind of vulnerabilities was proposed a hybrid process between the CERT-RMM and ISO/IEC 27002.

Finally, the proposed hybrid process will enable organizations to identify and manage vulnerabilities in their web applications; as future work the tool are being developed to automatize the proposed hybrid process to validate it.

References

1. Casaca, J.: Determinants of the information security effectiveness in small and medium sized enterprises. Proceedings in EIIC-The 3rd Electronic International Interdisciplinary Conference. pp. 495–500 (2014)
2. Kaspersky: Social engineering| internet security threats| kaspersky lab Mexico. http://latam.kaspersky.com/mx/internet-security-center/threats/malware-social-engineering (2015). Accessed 17 Jun 2015
3. Gartner: Gartner news room. http://www.gartner.com/newsroom/ (2014). Accessed 16 Feb 2015
4. NIST: News—NIST IT security. http://www.nist.org/news.php (2014). Accessed 16 Feb 2015
5. National Vulnerability Database: NVD—statistics search. https://web.nvd.nist.gov/view/vuln/statistics (2015). Accessed 16 Feb 2015
6. McAfee Labs: McAfee labs threats report, no. Nov 2014
7. OSVDB: OSVDB: Open Sourced Vulnerability Database. http://osvdb.org/ (2014). Accessed 07 Dec 2014
8. McAfee: McAfee labs informe sobre amenazas. (2014)
9. Kitchenham, B.: Evidence-based software engineering. Softw. Eng. (2004)
10. Hernández Saucedo, A.L.: Guía de ataques, vulnerabilidades, técnicas y herramientas para aplicaciones web. Recibe Revista Electrónica de Computación, biomédica y electrónica, no. 1, 2015
11. Singh, B., Kannojia, S.P.: A review on software quality models. 2013 Int. Conf. Commun. Syst. Netw. Technol., pp. 801–806 Apr 2013
12. Caralli, R., Allen, J., Curtis, P.: CERT® Resilience Management Model, v1. 0 (2011)
13. AENOR: UNE-ISO/IEC 27000 (2014)
14. ISO: ISO/IEC 27002:2013. https://www.iso.org/obp/ui/#iso:std:iso-iec:27002:ed-2:v1:en (2014). Accessed 16 Feb 2015
15. AENOR: UNE-ISO/IEC 27002 (2009)
16. ITGI: Alineando CobiT 4.1, ITIL V3, ISO/IEC 27002 en beneficio del negocio. (2008)
17. Madrid, E.P.: Sistemas y servicios digitales e híbridos de información. (2009)

Further Reading

18. W3af.org: w3af—Open Source Web Application Security Scanner. http://w3af.org/ (2015). Accessed 23 Jul 2015
19. Qualys: Qualys Web Application Scanning (WAS) | Qualys, Inc. https://www.qualys.com/enterprises/qualysguard/web-application-scanning/ (2014). Accessed 19 Dec 2014
20. Beyontrust: Web vulnerability management software | Assessment software, http://www.beyondtrust.com/Products/RetinaWebSecurityScanner/ (2014). Accessed 19 Dec 2014

Establishing the State of the Art of Frameworks, Methods and Methodologies Focused on Lightening Software Process: A Systematic Literature Review

Juan Miramontes, Mirna Muñoz, Jose A. Calvo-Manzano and Brisia Corona

Abstract There are several models used in the software process improvement such as CMMI and ISO/IEC 15504 that provide advantages to software organizations, like their improvement in ability and maturity which is reflected in their competitiveness. However, most of the time they are not properly implemented, demanding too many resources and long-term commitments, hindering its implementation in software organizations. Moreover, a properly implementation of a model or standard, involves not only the definition of processes, but introducing the organizations in the development of a continuous process improvement culture. In this context, a feasible way to achieve continuous process improvement is the optimization of processes through their lightening. This paper presents the results of a systematic review method in order to establish the state of art for lightening software process, focusing on three aspects: (1) frameworks, methods and methodologies; (2) targeted processes; and (3) strategies.

Keywords Lightening software process · Lightweight software process · Software process improvement · Systematic literature review

J. Miramontes (✉) · M. Muñoz · B. Corona
Centro de Investigación en Matematicas, Av. Universidad No. 222, 98068 Guadalupe, Zacatecas, Mexico
e-mail: juan.miramontes@cimat.mx

M. Muñoz
e-mail: mirna.munoz@cimat.mx

B. Corona
e-mail: brisia.corona@cimat.mx

J.A. Calvo-Manzano
Facultad de Informática Campus de Montegancedo S/N, Universidad Politécnica de Madrid, 28660 Madrid, Spain
e-mail: joseantonio.calvomanzano@upm.es

J. Mejia et al. (eds.), *Trends and Applications in Software Engineering*, Advances in Intelligent Systems and Computing 405,
DOI 10.1007/978-3-319-26285-7_7

1 Introduction

The software process improvement (SPI) aims to increase the efficient of software process in organizations, as well as the quality of software products through a continuous assess and adjustment of their software processes [1]. As a result, a set of process improvement models and standards have been proposed such as the Capability Maturity Model Integration (CMMI) (CMMI Product Team, [2], and the ISO/IEC 15504 [3] standard, which provide a set of best practices that has proved a high performance and successful in software organizations [4].

However, the goal of these models and standards are perceived for many organizations as a "heavyweight", that means too hard to understand and to implement [5, 6], because of the great amount of resources and the short and long-term commitments that they required. This makes difficult for small and medium enterprises to start and to carry out activities regarding the assessment and improvement due to the required time and costs [5–7].

In this context, it is important to emphasize that a proper implementation of a model or standard, involves not only the definition of processes, but it introduces to the organizations in the developing of a continuous improvement culture. Therefore, in organizations with defined processes based on the implementation of models and standards such as CMMI and ISO/IEC 15504 where they adopted a continuous improvement culture, optimizing their processes through their lightening is a feasible way. Quoting the phrase wrote by Saint-Exupery in [8]: "Perfection is achieved, not when there is nothing to add, but when there is nothing to take away".

This research aims to establish the state of the art regarding the lightening of software process focusing in three key aspects: frameworks, methods or methodologies used; targeted processes, and implemented strategies.

The rest of the paper is organized as follows: Sect. 2 shows the systematic review protocol developed for this research; Sect. 3 presents the analysis of the data extracted from the primary studies and; Sect. 4 summarizes the findings and outlines future work.

2 Systematic Literature Review

The Systematic Literature Review (SLR) is a method to identify, evaluate and interpret all relevant evidence available regarding a topic or research question. The SLR reduces the potential for bias in the search of studies because it has a defined protocol, which specifies the methods to be used and guides the systematic review. The SRL consisting of three main phases: planning the review, conducting the review, and reporting the results [9]. Next, the defined protocol for this research is briefly described.

2.1 Planning the Systematic Review

Planning is the first phase of the systematic review, it includes the following activities: identify the need to perform the review, specify the research questions, create the search string and select the data sources.

2.1.1 Identify the Need to Perform the SLR

One of the most important concerns in the software industry is the development of software products with the optimal use of resources, time and costs, i.e., a software organization needs to be efficient and have an optimal software development [10]. According to Rizwan & Hussain [11], for being efficient, an organization needs to lighter their software development processes. The systematic review is performed in order to know the current status of lightening software process, focusing on three key elements: used frameworks, methods and methodologies; targeted software process and; used strategies.

2.1.2 Specify the Research Questions

Three research questions were set: (RQ1) What frameworks, methods or methodologies exist for lightening or optimizing software process?; (RQ2) What are the most targeted software processes? and; (RQ3) What are the strategies used for lightening the software process?

2.1.3 Create the Search String

The words of the research questions considered key for the research were selected. As Table 1 shows, synonyms and terms related to the keywords were listed. Then, combining the keywords with the logical connectors such as "AND" and "OR" the search string was created.

2.1.4 Select the Data Sources

Data sources considered relevant in Software Engineering filed were selected as follows: (a) IEEE Xplore, (b) Elsevier Science (Science Direct), (c) SpringerLink and, (d) ACM Digital Library.

Table 1 Keywords and search string

Keywords	Synonyms or related words	Search string
Frameworks, methods, methodologies		(framework OR method OR methodology) AND (lightweight OR lean OR lighten OR light OR optimize OR optimizing) AND software process
Lightening	Lightweight/light-weight	
	Lean	
	Light	
	Lighten	
Optimize	Optimizing	
Software process		

2.2 *Conducting the Review*

The second phase of the SLR is conducting the review, which focuses on obtaining the primary studies. It includes the following activities: set the inclusion and exclusion criteria, select the primary studies, and data extraction.

2.2.1 Set the Inclusion and Exclusion Criteria

Inclusion criteria: (1) studies in the English or Spanish languages; (2) studies between the years 2008 and 2015; (3) studies containing at least two keywords in the title and abstract; (4) studies containing the analysis, evaluation and application of models or software process standards; (5) studies containing case studies or validated proposals; and (6) studies showing the results of software processes lightening or optimization.

Exclusion criteria: (1) studies that do not contain information about software process lightening of software processes or lightweight software process; (2) studies repeated in more than one source; (3) Inaccessible studies.

2.2.2 Select the Primary Studies

The primary studies were obtained by executing the selection process. It consists of five steps: (1) take the search string and adapt it to the data sources of the search engine; (2) filter studies using the first three inclusion criteria; (3) read the titles and abstracts to identify potentially relevant studies; (4) apply the remaining inclusion and exclusion criteria on reading the introduction, methods, conclusion and, if necessary, all the study; and (5) select the primary studies.

Figure 1 shows the implementation of the steps and the number of studies obtained. As figure shows, of 562,256 studies found running the search string in the data sources, only 32 studies met all the inclusion and exclusion criteria. These

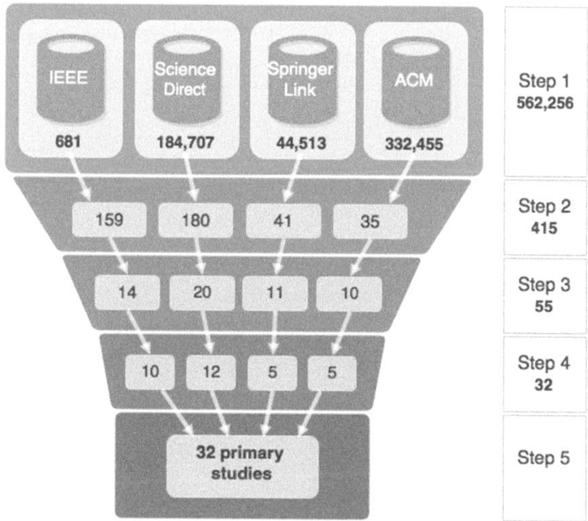

Fig. 1 Primary studies obtained performing the selection process

studies were analyzed and used for this research. The list of primary studies is shown in Appendix A.

2.2.3 Data Extraction

The data extracted from the primary studies was registered in a developed template in a spreadsheet editor (Microsoft Excel™), which contains the following data: title, author, year, keywords, data source, goal, problem, strategy, validation, findings, targeted phase o process, and proposal type (method, methodology or framework).

3 Results Analysis

This section shows the analysis of the main results.

3.1 Frameworks, Methods and Methodologies

This analysis aims to identify those frameworks, methods and methodologies developed in order to lighten software process. As Fig. 2 shows, not only frameworks, methods and methodologies were found; in addition, other proposals such as tools and processes were gotten. The results show that most of the proposals are focused on the development of frameworks and tools, followed by methods and methodologies.

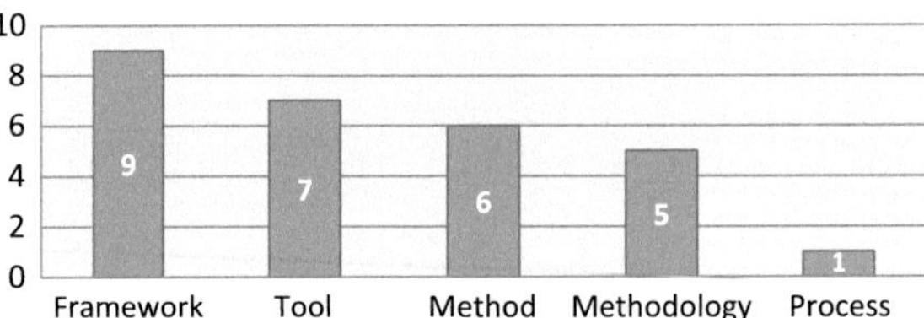

Fig. 2 Number of studies by proposal type

The Appendix B includes a summary of the identified proposals, including information of their goal and type of proposal.

3.2 Targeted Processes

This analysis aims to identify targeted processes for their lightening. The results of this analysis shows that most of the targeted processes are related to process areas of the Capability Maturity Model Integration for development (CMMI-DEV) and the life cycle development stages. Therefore, processes and life cycle stages were classified within the process areas of CMMI-DEV. To perform the classification, studies that lighten phases such as design and software architecture were considered in the technical solution process area, and software testing were considered in the validation process area or verification as appropriate. The number of studies classified by process areas and maturity levels are shown in Table 2.

As shown in Table 2, the process areas most targeted are technical solution (5), validation (5), organizational process focus (4), and requirements development (3). Moreover, it can be observed that process areas of maturity level 3 are the most focused.

Besides, this section includes an analysis that classifies the targeted process according to the categories proposed by CMMI-DEV: project management, support, process management and engineering (see Fig. 3). The results show that most of the targeted processes are within the engineering and project management categories.

3.3 Used Strategies for Software Process Lightening

A categorization was made to know the strategies used for lightening software process. The classification includes the following strategies: the use of tools; the combination of formal and agile software development (ASD) practices; the identification of success factors in agile development; the identification of the organizational best practices; the use of meetings among stakeholders; the design of

Table 2 Number of studies by process area

Maturity level 2	Abbr.	SN[a]	Maturity level 3	Abbr.	SN[a]
Configuration management	CM	2	Decision analysis and resolution	DAR	1
Measurement and analysis	MA	2	Integrated project management	IPM	1
Project monitoring and control	PMC	2	Organizational process definition	OPD	2
Project planning	PP	2	Organizational process focus	OPF	4
Process and product quality assurance	PPQA	2	Organizational training	OT	1
Requirements management	REQM	2	Product integration	PI	2
Supplier agreement management	SAM	2	Requirements development	RD	3
			Risk management	RSKM	1
			Technical solution	TS	5
			Validation	VAL	5
			Verification	VER	2
	Total	14		Total	27
Maturity level 4	Abbr.	SN[a]	Maturity level 5	Abbr.	SN[a]
Organizational process performance	OPP	0	Causal analysis and resolution	CAR	1
Quantitative project management	QPM	0	Organizational performance management	OPM	0
	Total	0		Total	1

[a]SN = Number of studies

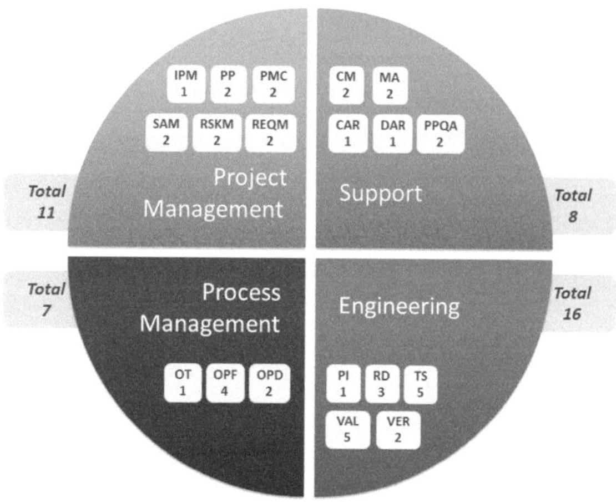

Fig. 3 Number of process areas by CMMI categories

Table 3 Strategies for lightening software process

Strategy	Description	SN[a]
Use of tools	Use of tools for automating the process, reducing the time and effort for its implementation.	8
Combination of formal practices and ASD	Integrate practices of the formal methods with the agile software development practices.	6
Identifying success factors in ASD	Identify critical elements of success in agile methodologies and modeling a software process.	3
Identification of organizational best practices	Identify key practices that an organization needs to achieve its business objectives and define the process based on them.	3
Meetings among stakeholders	Streamline the outputs of a process by conducting meetings or workshops with stakeholders.	3
Design methodology	Designing a methodology for defining lightweight software processes.	3
Use of Lean methods	Using Lean processes: eliminate activities that do not add value to the process.	3

[a]SN = Number of studies

methodologies to lighten software process; and the use of Lean methods. Table 3 shows the strategies classification including a description of the strategy and the number of found studies.

According to the obtained results, "the use of tools" is the most used strategy for lightening software process. However, tools are focused on the process automation, reducing the time and effort that takes its execution, but they do not analyze the process.

Moreover, due to the nature of the strategies, it is identified that the strategies are not mutually exclusive. Then, they can be integrated or combined for lightening software processes.

4 Conclusions and Future Work

A systematic literature review was conducted to establish the state of the art of lightening software process focus on three key elements: frameworks, methods and methodologies, targeted software process and strategies. After applying an inclusion and exclusion criteria, from 562,256 only 32 studies were selected as "primary studies".

The main results shows that the model focused for lightening processes is the Capability Maturity Model Integration for Development (CMMI-DEV); evidence of proposals using the ISO 15504 standard was not found.

Other found results are that the most targeted processes are technical solution and software validation, followed by the organizational process focus, and requirements development. Besides, most of the targeted processes are included in the maturity level 2 and 3, and the engineering and project management category.

Finally, the most used strategies for lightening software process are based on the use of tools; however, these tools are focused on reducing the time and effort for the process execution, not performing an analysis to optimize the process activities. Furthermore, analyzing other strategies it was concluded that they could be integrated or combined to analyze the processes for their proper lightening.

Based on the above-mentioned as future work, the goal of this research is to develop a method for lightening software process, analyzing the use of some of the identified strategies. The proposed method will aim to support software organizations in a continuous improvement optimizing their process through lightening them without losing the requirements to achieve a certification on models or standards such as CMMI or ISO 15504.

Appendix A: Primary Studies

S1 Akbar, R., Hassan, M. F., & Abdullah, A. (2012). A framework of software process tailoring for small and medium size IT companies. *2012 International Conference on Computer & Information Science (ICCIS),* 2, 914–918. http://doi.org/10.1109/ICCISci.2012.6297156

S2 Al-Tarawneh, M. Y., Abdullah, M. S., & Ali, A. B. M. (2011). A proposed methodology for establishing software process development improvement for small software development firms. *Procedia Computer Science, 3,* 893–897. http://doi.org/10.1016Zj.procs.2010.12.146

S3 Alwardt, A. L., Mikeska, N., Pandorf, R. J., & Tarpley, P. R. (2009). A lean approach to designing for software testability. *AUTOTESTCON (Proceedings),* 178–183. http://doi.org/10.1109/AUTEST.2009.5314039

S4 Camargo, K. G., Ferrari, F. C., & Fabbri, S. C. (2015). Characterising the state of the practice in software testing through a TMMi-based process. *Journal of Software Engineering Research and Development,* 3(1), 7. http://doi.org/10.1186/s40411-015-0019-9

S5 Edison, H., Wang, X., & Abrahamsson, P. (2015). Lean startup: Why Large Software Companies Should Care. *Scientific Workshop Proceedings of the XP2015 on - XP '15 Workshops,* 1–7. http://doi.org/10.1145/2764979.2764981

S6 Farid, W. M. (2012). The Normap methodology: Lightweight engineering of non-functional requirements for agile processes. *Proceedings - Asia-Pacific Software Engineering Conference, APSEC, 1,* 322–325. http://doi.org/10.1109/APSEC.2012.23

S7 Farrow, A., & Greene, S. (2008). Fast & predictable—A lightweight release framework promotes agility through rhythm and flow. *Proceedings—Agile 2008 Conference,* 224–228. http://doi.org/10.1109/Agile.2008.83

S8 Funkhouser, O., Etzkorn, L. H., & Hughes, W. E. (2008). A lightweight approach to software validation by comparing UML use cases with internal program documentation selected via call graphs. *Software Quality Journal, 16*(1), 131–156. http://doi.org/10.1007/s11219-007-9034-3

S9 Garcia, J., Amescua, A., Sanchez, M. I., & Bermon, L. (2011). Design guidelines for software processes knowledge repository development. *Information and Software Technology, 53*(8), 834–850. http://doi.org/10.1016/j.infsof.2011.03.002

S10 Ivarsson, M., & Gorschek, T. (2012). Tool support for disseminating and improving development practices. *Software Quality Journal,* 20(1), 173–199. http://doi.org/10.1007/s11219-011-9139-6

S11 Kim, T., Chandra, R., & Zeldovich, N. (2013). Optimizing unit test execution in large software programs using dependency analysis. *Proceedings of the 4th Asia-Pacific Workshop on Systems—APSys '13,* 1-6. http://doi.org/10.1145/2500727.2500748

S12 Kirk, D., & Tempero, E. (2012). A lightweight framework for describing software practices. *Journal of Systems and Software,* 85(3), 582–595. http://doi.org/10.1016/j.jss.2011.09.024

S13 Kruchten, P. (2011). A plea for lean software process models. *Procs. 2011 International Conference on Software and Systems Process (ICSSP), 1* (604), 235–236. http://doi.org/10.1145/1987875.1987919

S14 Lehtinen, T. O. a, Mantyla, M. V., & Vanhanen, J. (2011). Development and evaluation of a lightweight root cause analysis method (ARCA method)—Field studies at four software companies. *Information and Software Technology,* 53(10), 1045–1061. http://doi.org/10.1016/j.infsof.2011.05.005

S15 Lin, W. L. W., & Fan, X. F. X. (2009). Software Development Practice for FDA-Compliant Medical Devices. *2009 International Joint Conference on Computational Sciences ana Optimization, 2,* 388–390. http://doi.org/10.1109/CS0.2009.191

S16 Misra, S. C., Kumar, V., & Kumar, U. (2009). Identifying some important success factors in adopting agile software development practices. *Journal of Systems and Software,* 82(11), 1869–1890. http://doi.org/10.1016/j.jss.2009.05.052

S17 Motta, A., & Mangano, N. (2013). Lightweight Analysis of Software Design Models at the Whiteboard, 18–23.

S18 Pang, H., Zhou, L., & Chang, X. (2011). Lightweight web framework oriented on page flow component. *Proceedings 2011 International Conference on Mechatronic Science, Electric Engineering and Computer, MEC 2011,* 1248–1251. http://doi.org/10.1109/MEC.2011.6025694

S19 Park, S., & Bae, D.-H. (2011). An approach to analyzing the software process change impact using process slicing and simulation. *Journal of Systems and Software,* 84(4), 528–543. http://doi.org/10.1016/j.jss.2010.11.919

S20 Peng, X., Chen, B., Yu, Y., & Zhao, W. (2012). Self-tuning of software systems through dynamic quality tradeoff and value-based feedback control loop. *Journal of Systems and Software,* 85(12), 2707–2719. http://doi.org/10.1016/jJss.2012.04.079

S21 Petersen, K., & Wohlin, C. (2010). Software process improvement through the Lean Measurement (SPI-LEAM) method. *Journal of Systems and Software,* 83(7), 1275–1287. http://doi.org/10.1016/j.jss.2010.02.005

S22 Pettersson, F., Ivarsson, M., Gorschek, T., & Ohman, P. (2008). A practitioner's guide to light weight software process assessment and improvement planning. *Journal of Systems and Software,* 81(6), 972–995. http://doi.org/10.1016/jJss.2007.08.032

S23 Pino, F. J., Pedreira, O., Garcia, F., Luaces, M. R., & Piattini, M. (2010). Using Scrum to guide the execution of software process improvement in small organizations. *Journal ol Systems and Software,* 83(10), 1662–1677. http://doi.org/10.1016/jJss.2010.03.077

S24 Rapp, D., Hess, A., Seyff, N., Sporri, P., Fuchs, E., & Glinz, M. (2014). Lightweight Requirements Engineering Assessments in Software Projects, 354–363.

S25 Rigby, P., Cleary, B., Painchaud, F., Storey, M. A., & German, D. (2012). Contemporary peer review in action: Lessons from open source development. *IEEE Software,* 29(6), 56–61. http://doi.org/10.1109/MS.2012.24

S26 Rodriguez, P., Mikkonen, K., Kuvaja, P., Oivo, M., & Garbajosa, J. (2013). Building lean thinking in a telecom software development organization: strengths and challenges. *Proceedings of the 2013 International Conference on Software and System Process—ICSSP 2013,* 98. http://doi.org/10.1145/2486046.2486064

S27 Rubin, E., & Rubin, H. (2011). Supporting agile software development through active documentation. *Requirements Engineering,* 16(2), 117–132. http://doi.org/10.1007/s00766-010-0113-9

S28 Selleri, F., Santana, F., Soares, F., Lima, A., Monteiro, I., Azevedo, D., … Meira, D. L. (2015). Using CMMI together with agile software development: A systematic review. *Information and Software Technology, 58,* 20–43. http://doi.org/10.1016/jjnfsof.2014.09.012

S29 Stankovic, D., Nikolic, V., Djordjevic, M., & Cao, D.-B. (2013). A survey study of critical success factors in agile software projects in former Yugoslavia IT companies. *Journal of Systems and Software,* 86(6), 1663–1678. http://doi.org/10.1016/jJss.2013.02.027

S30 Vale, T., Cabral, B., Alvim, L., Soares, L., Santos, A., Machado, I., … Almeida, E. (2014). SPLICE: A Lightweight Software Product Line Development Process for Small and Medium Size Projects. *2014 Eighth Brazilian Symposium on Software Components, Architectures and Reuse,* 42–52. http://doi.org/10.1109/SBCARS.2014.11

S31 Vanhanen, J., Mantyla, M. V., & Itkonen, J. (2009). Lightweight elicitation and analysis of software product quality goals—A multiple industrial case study. *2009 3rd International Workshop on Software Product Management, IWSPM 2009,* 27–30. http://doi.org/10.1109/IWSPM.2009.5

S32 Zarour, M., Abran, A., Desharnais, J.-M., & Alarifi, A. (2015). An investigation into the best practices for the successful design and implementation of lightweight software process assessment methods: A systematic literature review. *Journal of Systems and Software, 101,* 180–192. http://doi.org/10.1016/jJss.2014.11.041

Appendix B: Proposals Type and Their Goal

ID	Proposal type	Authors	Goal
S1	Framework	(Akbar, Hassan, & Abdullah, 2012)	Presents a meta-model framework for adaptation lightweight software processes.
S2	Methodology	(Al-Tarawneh, Abdullah, & Ali, 2011)	Presents a methodology for SPI in SMEs using CMMI.
S3	Framework	(Alwardt, Mikeska, Pandorf, & Tarpley, 2009)	Shows how Lean 123 with an automatic approach in software testing saves costs and improves product quality.
S4	Tool	(Camargo, Ferrari, & Fabbri, 2015)	Identifies a set of key practices to support a generic lightweight process for software testing based on TMMi.
S5	Method	(Edison, Wang, & Abrahamsson, 2015)	Analyses why large companies should adopt Lean Startup to seek radical innovation.
S6	Methodology	(Farid, 2012)	Presents NORMAP, a lightweight methodology for non-functional requirements in agile process.
S7	Framework	(Farrow & Greene, 2008)	Presents a lightweight framework for the software release, in order to optimize time and delivering high quality to customers.
S8	Methodology	(Funkhouser, Etzkorn, & Hughes, 2008)	Presents a methodology to automate software validation.
S9	Tool	(García, Amescua, Sánchez, & Bermón, 2011)	Design guidelines for implementing a Process Asset Library (PAL) via a wiki, for storing organizational best practices.
S10	Tool	(Ivarsson & Gorschek, 2012)	Presents a tool support for disseminating and improving practices used in an organization based on the Experience Factory approach.
S11	Tool	(Kim, Chandra, & Zeldovich, 2013)	Demonstrates that TAO tool can reduce unit test execution time in two large Python software projects by over 96 %.
S12	Framework	(Kirk & Tempero, 2012)	Develop a framework to capture the best practices of successful companies.

(continued)

(continued)

ID	Proposal type	Authors	Goal
S13	Other[a]	(Kruchten, 2011)	Explain why large and heavy processes are not suitable for software development and should be used lightweight processes.
S14	Method	(Lehtinen, Mäntylä, & Vanhanen, 2011)	Presents a lightweight method for RCA (Root Cause Analysis) called ARCA, in which the detection of a problem is based on a group meeting focused on the issue.
S15	Methodology	(Lin & Fan, 2009)	Shows the practice to develop software for medical devices that need formal and rigorous processes using a hybrid approach between CMMI and agile software development (ASD).
S16	Framewok	(Misra, Kumar, & Kumar, 2009)	Presents the state of the art on the identification of success factors in adopting agile software development practices.
S17	Tool	(Motta & Mangano, 2013)	Improves the Calico tool to allow a lightweight analysis, giving rapid feedback to the developer when developing a design.
S18	Framework	(Pang, Zhou, & Chang, 2011)	Designs a lightweight web framework based on page flow to improve software development and reduce development costs.
S19	Tool	(Park & Bae, 2011)	Proposes an approach for analyzing the impact of the change of a software process using slicing y simulation.
S20	Method	(Peng, Chen, Yu, & Zhao, 2012)	Proposes a method of selftuning that can dynamically capture the quality requirements and make tradeoff decisions through a Preference-Based Goal Reasoning procedure.
S21	Method	(Petersen & Wohlin, 2010)	Proposes a novel approach to bring together the quality improvement paradigm and lean software development practices, called SPI-LEAM method.
S22	Framework	(Pettersson et al., 2008)	Presents a guide to light weight software process assessment and improvement planning.
S23	Framework	(Pino, Pedreira, García, Luaces, & Piattini, 2010)	Proposes a "Lightweight process to incorporate improvements", using the philosophy of the Scrum agile method, aiming to give detailed guidelines for incorporating process improvements in small companies.

(continued)

(continued)

ID	Proposal type	Authors	Goal
S24	Method	(Rapp et al., 2014)	Develops a lightweight method to answer questions related to the quality of requirements engineering process, so that a company can be assessed and improved.
S25	Tool	(Rigby, Cleary, Painchaud, Storey, & German, 2012)	Describes lessons learned from the review process code in OSS (Open Source Software) to transfer them to development of proprietary software.
S26	Methodology	(Rodríguez, Mikkonen, Kuvaja, Oivo, & Garbajosa, 2013)	Explores how Lean principles are implemented in the software development companies and the challenges in the implementation of Lean.
S27	Framework	(Rubin & Rubin, 2011)	Proposes the system design Active Documentation Software Design (ADSD), by this, source code incorporates documentation statements.
S28	Other[a]	(Selleri et al., 2015)	Evaluates, synthesize, and present results on the use of the CMMI in combination with agile software development, and thereafter to give an overview of the topics researched.
S29	Other[a]	(Stankovic, Nikolic, Djordjevic, & Cao, 2013)	Presents the results of an empirical study to determine the critical factors that influence the success of agile projects.
S30	Method	(Vale et al., 2014)	Introduces SPLICE, a lightweight process that combines agile development practices with Software Product Line Engineering.
S31	Method	(Vanhanen, Mäntylä, & Itkonen, 2009)	Presents a method that gathers relevant stakeholders to elicit, prioritize, and elaborate the quality goals of a software product
S32	Other[a]	(Zarour, Abran, Desharnais, & Alarifi, 2015)	Presents the results of performing a systematic literature review focused on the best practices that help to SPA (Software Process Assessment) researchers and practitioners in designing and implementing lightweight assessment methods.

[a]In these studies the subject of lightening software processes are analyzed, but they do not present a proposal

References

1. Petersen, K., Wohlin, C.: Software process improvement through the Lean Measurement (SPI-LEAM) method. J. Syst. Softw. **83**(7), 1275–1287 (2010). doi:10.1016/j.jss.2010.02.005
2. CMMI Product Team: CMMI® for Development, Version 1.3. Pittsburgh, PA (2010)
3. ISO/IEC: ISO/IEC 15504 Information Technology—Process Assessment (Parts 1–5) (2004)
4. Pettersson, F., Ivarsson, M., Gorschek, T., Öhman, P.: A practitioner's guide to light weight software process assessment and improvement planning. J. Syst. Softw. **81**(6), 972–995 (2008). doi:10.1016/j.jss.2007.08.032
5. Kuilboer, J., Ashrafi, N.: Software process and product improvement: an empirical assessment. Inf. Softw. Technol. **42**(1), 27–34 (2000). doi:10.1016/S0950-5849(99)00054-3
6. Reifer, D.J.: The CMMI: it's formidable. J. Syst. Softw. **50**(2), 97–98 (2000). doi:10.1016/S0164-1212(99)00119-3
7. Villalón, J.A.C.M, Agustín, G.C., Gilabert, T.S.F., Seco, A.D.A., Sánchez, L.G., Cota, M.P: Experiences in the application of software process improvement in SMES. Softw. Qual. J. **10**, 261–273 (2002)
8. Kruchten, P.: A plea for lean software process models. In: Proceedings 2011 International Conference on Software and Systems Process (ICSSP), vol. 1, no.604, pp. 235–236 (2011). http://doi.org/10.1145/1987875.1987919
9. Selleri, F., Santana, F., Soares, F., Lima, A., Monteiro, I., Azevedo, D., Meira, D.L.: Using CMMI together with agile software development: a systematic review. Inf. Softw. Technol. **58**, 20–43 (2015). doi:10.1016/j.infsof.2014.09.012
10. Garzás, J., Pino, F.J., Piattini, M., Fernández, C.M.: A maturity model for the Spanish software industry based on ISO standards. Comput. Stand. Interfaces **35**(6), 616–628 (2013). doi:10.1016/j.csi.2013.04.002
11. Qureshi, M.R.J., Hussain, S.A.: An adaptive software development process model. Adv. Eng. Softw. **39**(8), 654–658 (2008). doi:10.1016/j.advengsoft.2007.08.001

Structure of a Multi-model Catalog for Software Projects Management Including Agile and Traditional Practices

Andrés Felipe Bustamante, Jesús Andrés Hincapié and Gloria Piedad Gasca-Hurtado

Abstract Software development projects can be managed under a great variety of methodologies and frameworks. Use of traditional frameworks can lead to extended planning stages that take a significant amount of time. Agile methodologies are designed to accelerate the creation of value through an incremental evolutionary process, where activities that create more costumer value are prioritized. Use of agile methodologies can relegate important factors in project management, if they are not included in the planning phase, since these methodologies do not propose dimensions that traditional methodologies do. We pretend to identify whether it is possible to combine agile methodologies and traditional models to define a catalog of best practices for software development project planning. We propose to reduce the complexity of implementing an integrated agile/traditional model, through a methodological catalog for project planning. We present a high level solution design, including initial concepts of the model for a future catalog development.

Keywords Process improvement · Software project management · Agile methodologies · Traditional reference frameworks · Multi-model environment

1 Introduction

When planning and executing software development projects, we can take into account a great variety of methodologies and reference frameworks that provide the foundations for such projects, as it is illustrated by [1].

A.F. Bustamante · J.A. Hincapié (✉) · G.P. Gasca-Hurtado
Universidad de Medellín, Medellín, Colombia
e-mail: jehincapie@udem.edu.co; jahlon@gmail.com

A.F. Bustamante
e-mail: estmae_ingsw@udem.edu.co

G.P. Gasca-Hurtado
e-mail: gpgasca@udem.edu.co

J. Mejia et al. (eds.), *Trends and Applications in Software Engineering*,
Advances in Intelligent Systems and Computing 405,
DOI 10.1007/978-3-319-26285-7_8

In this paper we study two groups of reference frameworks. In the one hand, we have the traditional reference frameworks such as PMI [2], CMMI [3], ISO 9001 [4] or TSP [5]. In the other hand, we have agile methodologies such as Scrum [6], Kanban [7] or XP [8].

Traditional reference frameworks for project management are based on a set of best practices and formats that help considering important aspects when managing a project. Such aspects must be defined from the beginning of the project with reference frameworks such as CMMI, which gives guidelines for developing effective processes and provides "a set of best practices to develop products and services, covering the whole lifecycle of the project" [9].

The use of traditional reference frameworks can lead to extended planning stages, where taking into account all the relevant project aspects can take a considerable amount of time. This can impact development time negatively, since more documentation detail is required and more time has to be invested in process improvement activities [10].

Agile methodologies propose frameworks to accelerate the generation of value through an incremental evolutionary process, where activities that create more costumer value are prioritized. This requires a more active participation of the client and a relentless decision making in order to define what is to be implemented in each process iteration [10].

Agile methodologies "propose to deliver results in small periods of time, which required great discipline" [9], however they can also relegate important factors of project planning, since this methodologies do not consider the dimensions that traditional reference frameworks do. In this sense, it is not easy for organizations with an agile culture to be rigorous with the requirements of traditional reference frameworks such as CMMI or ISO [10].

This work aims at proposing a solution to reduce the complexity involved in implementing an integrated agile/traditional model, through a methodological catalog for project planning, establishing execution paths guided by dimensions defined within the catalog. To this end, we pretend to analyze studies about software development that consider best practices in agile methodologies and traditional reference frameworks regarding project management.

This problem has been addressed in other works, using knowledge management strategies by means of reusable project patterns [11, 12]. The idea is to encapsulate information about the recommended practices to develop specific software projects.

Although our proposal also seeks to combine practices from different methodologies or reference frameworks, its difference lies on the fact that we intent to provide a guideline in terms of dimensions that take into account specific practices from the agile and the traditional world.

This paper is structured as follows. In Sect. 2 we present the characteristics of the performed study about agile and traditional approaches to software project management, and analyze the results of the study discussing the advantages and disadvantages of the reviewed works. In Sect. 3 we present the proposal of the multi-model catalog for software project management including agile practices. In Sect. 4 we discuss our proposal and draw some conclusions and future work.

2 Study of Agile and Traditional Approaches to Software Project Management

For our proposal, we performed a study of several works that consider approaches to software project management both from the agile and the traditional perspective. We define the following characteristics based on the protocol established in [13].

The first characteristic was the focus of the question, which was to identify solutions that take into account best practices in agile methodologies and traditional reference frameworks regarding software development projects management.

Secondly, we define the problem as follows: selecting a software project planning and execution methodology from a traditional reference framework or an agile methodology exclusively can generate very robust planning stages, where taking into account all the relevant project aspects can take a considerable amount of time, or where not all the relevant project aspects are considered.

The third characteristic was the questions to guide the search process. We define two questions for the study:

1. Which solutions have been proposed in recent years that consider best practices of agile methodologies and traditional reference frameworks for software development project management?
2. What are the advantages and disadvantages that have been identified when using best practices of agile methodologies and traditional reference frameworks for software development project management?

Finally, we define the effect we wanted from the study, which was to identify best practices of software development project management from the integration of agile methodologies and traditional reference frameworks.

We conducted a search in several sources looking for works related to software project management with traditional reference frameworks and agile methodologies. After a review process, we selected several works that are listed in Table 1.

The selected works were evaluated according to the following four criteria:

CRIT1 was whether the work poses an approach to software development project management from the perspective of traditional reference frameworks.
CRIT2 was whether the work poses and approach to software development project management from the perspective of agile methodologies.
CRIT3 was whether the work integrates or compares traditional reference frameworks and agile methodologies in order to abstract best practices of software projects management.
CRIT4 was whether the work identifies advantages, problems or improvement opportunities regarding software development projects management.

The criteria were measured with the following scale: 0 means the work does not meet the criterion; ½ means the work partially meets the criterion; and 1 means the work completely meets the criterion. Results are shown in Table 2.

Table 1 Selected works

ID	Works	Keywords
PAP01	Mixed agile/traditional project management methodology—reality or illusion? [14]	Agile project management; traditional project management; methodology
PAP02	Dealing the selection of project management through hybrid model of verbal decision analysis [15]	Verbal Decision Analysis; ZAPROS III-i; ORCLASS; Project Management; Specific Practices; CMMI
PAP03	The root cause of failure in complex IT projects: complexity itself [16]	project management; project failure; information technology; complex adaptive systems; PMBOK; agile
PAP04	Barriers towards integrated product development—challenges from a holistic project management perspective [17]	Integrated product development; Project management; Project governance; Case study method
PAP05	The integration of project management and organizational change management is now a necessity [18]	Organizational change management; Project management; Organizational change; Organization
PAP06	A reduced set of RUP roles to small software development teams [19]	RUP; small teams; SME; RUP tailoring
PAP07	A scrum-based approach to CMMI maturity level 2 in web development environments [9]	CMMI (Capability Maturity Model Integration), Agile methodologies, Scrum, Web Engineering
PAP08	Light maturity models (LMM): an Agile application [20]	Agile, Maturity Models, Crosby, Continual Process Improvement
PAP09	Agility at scale: economic governance, measured improvement, and disciplined delivery [21]	Software process improvement, agile development, integration first, economic governance, measured improvement
PAP10	Combining maturity with agility: lessons learnt from a case study [10]	Management, Measurement, Design, Documentation, Experimentation, Human Factors, Standardization, Verification
PAP11	Theory based software engineering with the SEMAT kernel: preliminary investigation and experiences [22]	Management—lifecycle, productivity, programming teams, software process models
PAP12	A pattern system of underlying theories for process improvement [23]	Productivity, Software Process Models Economics, Management, Theory, Process improvement, best practices

After the assessment, we could observe, in general terms, that the evaluation criteria were met. Works PAP01–PAP07, PAP10 and PAP12 deeply explore traditional reference frameworks regarding projects management, giving a thorough description of the approach and recommending best practices for software development projects.

Špunda poses a clear analysis, identifying the basic idea behind traditional approaches, where "projects are expected to be relatively simple, predictable and linear, with clearly defined limits that allow detailed planning and execution without many changes" [14].

Table 2 Evaluation of works

		Evaluation criteria			
		CRIT1	CRIT2	CRIT3	CRIT4
Works	*PAP01*	1	1	1	1
	PAP02	1	1	1	1
	PAP03	1	1	1/2	1/2
	PAP04	1	1/2	0	1/2
	PAP05	1	1	1	1
	PAP06	1	1	1	1
	PAP07	1	1	1	1
	PAP08	1/2	1/2	1	1/2
	PAP09	1/2	1	1	1
	PAP10	1	1	1	1
	PAP11	1/2	1	1/2	1
	PAP12	1	1	1	1

Definitions such as the above are common among different authors, which explain projects, following a traditional framework, as "rational systems where artifacts resulting from the planning and design stages are a formalized structure to be executed with authority by the project manager" [16].

Agile methodologies analysis is presented in all the works evaluated, in greater depth in papers PAP01–PAP03, PAP05–PAP07 and PAP09–PAP12, where they explain the importance of including agile practices in project planning, emphasizing in software development projects.

The papers coincide in the idea that "agile project planning is well aware of the necessity of including concerns about social systems, focusing on the importance of culture, people development, self-management, self-discipline, participative decision making, focus on user and less bureaucracy" [18].

They also focus on risk control through agile methodologies, by posing software development in small periods of time and including customer participation in the decision making of each project stage [10].

Integration of traditional reference frameworks and agile methodologies is considered, starting from recommended best practices from both perspectives.

Works PAP01, PAP02, PAP05–PAP10 and PAP12 present a hybrid integrated model that includes practices from agile methodologies as well as traditional reference frameworks. All others works, except PAP03, PAP04 and PAP08 present problems to explain why it is useful to include best practices of agile methodologies and traditional frameworks.

Considering all these works, we can say that there is a general idea of having different methodologies combined in order to improve planning, taking into account integration of agile practices and traditional practices as it is explained by

Torrecillas Salinas et al., where traditional reference frameworks contribute to establish what needs to be done, and agile methodologies state how to do it [9].

The analysis of the selected works showed that there are clear advantages when integrating agile methodologies and traditional reference frameworks, as well as there are some opportunities for improvement.

Among the advantages that became evident, we can highlight some that come from the agile methodologies such as quick feedback, greater tolerance for change, risks management [10] and a social approach that empowers clients and project members [18]; we can also highlight advantages that come from traditional reference frameworks such as having a good project structure, having clear and precise artifacts [14] and understanding key activities and sets of best practices to fully cover the project life cycle [9].

3 Multi-model Catalog for Software Project Management

Organizations are using a growing set of international standards and models to manage their businesses, increase customer satisfaction, attain competitive advantage, and achieve process performance and regulatory compliance [24]. For this reason, it is likely that a company uses more than one process improvement model, standard and technology [25]. Multi-model Environment is an approach of the Software Engineering Institute developed to harmonize the process improvement models [26].

So far, we have shown that the integration between agile methodologies and traditional reference frameworks is a reality that is being contemplated in the industry today, with approaches that analyze mixed solutions, abstract models and sets of best practices of both sides.

However, when a company or a team is about to start a new project, and wants to tackle it from a mixed agile/traditional angle, it is required a deep analysis of topics regarding the company environment and the lack of clarity when several models are integrated. The idea is to reduce the risk of redundant or unproductive tasks and provide tools to determine what to use and how to use it properly.

The proposed solution to this problem is to include activities homologation catalogs between different methodologies. Such catalogs will consider practices categorization by framework in order to guide the process of deciding what to use and how to use it to assemble a self-defined methodology, according to the specific needs of a company and regardless of the model or standard used in it.

Starting from company needs and characteristics, and including agile and traditional reference frameworks, we propose four dimensions that help to structure the company's integrated methodological framework and enable it to start from an established path. An approach of this catalog can be seen in Fig. 1.

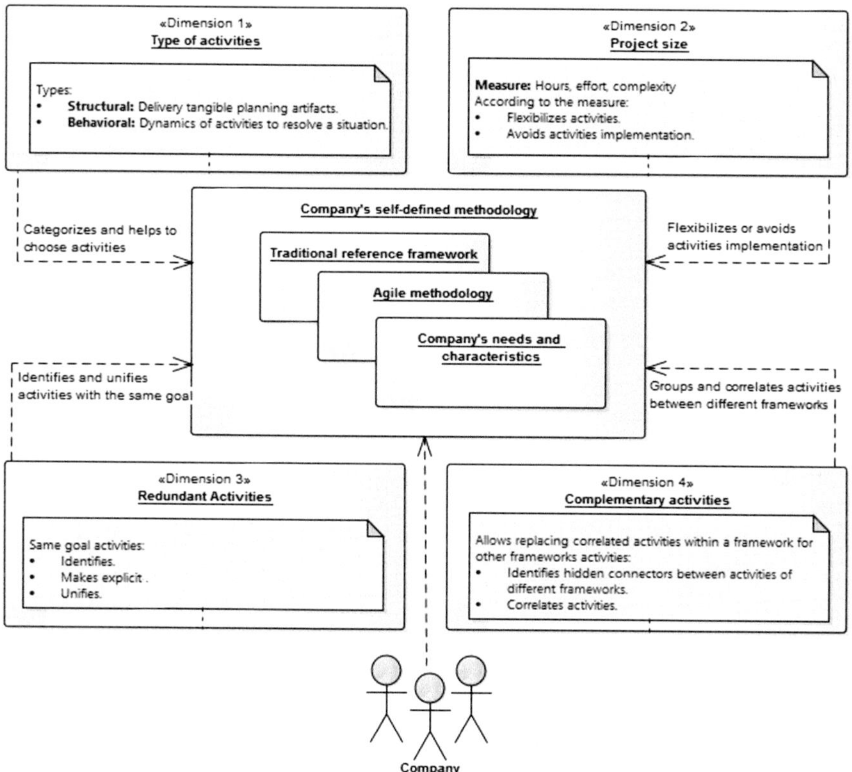

Fig. 1 Multi-model catalog for software project management

3.1 Dimension 1—Type of Activities

This dimension categorizes activities in 2 different types:

- Structural: Delivery tangible planning artifacts.
- Behavioral: Dynamics of activities to resolve a situation.

Similar to catalogs of other areas like Gamma et al. present in "Design Patterns: Elements of Reusable Object-Oriented Software" [27], also known as GoF patterns, the initial point to consider in the proposed catalog is a categorization under which activities are classified.

For software development project planning and its implementation, it is possible to start with these two types of activity, where the former gives guidelines of tangible artifacts in the planning stages, and the latter gives dynamic patterns of activities to solve specific situations.

3.2 Dimension 2—Project Size

This dimension adjusts or avoids activities implementation, according to the project size, which could be measured by hours, effort or complexity.

It is necessary to set an importance factor for activities, according to the project size, so it is possible to avoid non-essential activities or execute activities in a more agile way than defined by a framework.

3.3 Dimension 3—Redundant Activities

This dimension identifies and unifies activities with the same goal, making explicit the activities and its focus.

When combining methodologies, there is a risk of doing activities of different frameworks that can be focused on the same goal, but have significant differences in the methodological framework to which they belong, where it is not so explicit that they are pursuing the same goal.

Activities homologation between different methodological frameworks, including information about activities usefulness within a project, is the cornerstone of a catalog that really guides its users in the selection of the best way to go.

3.4 Dimension 4—Complementary Activities

This dimension groups and correlates activities between different frameworks, identifying connectors of activities that do not belong to the same framework.

It is a fact that in any methodology there are activities that depend on others, however, when methodologies are combined, it is likely that some dependencies between activities of the same methodological framework disappear, because such activities are replaced by others of a different framework.

Here, the idea is that replacement activities of a different framework allow reaching the objective that original dependencies sought.

Each dimension arises from the need to simplify the inclusion analysis of an activity or set of activities in a customized approach, based on the integration of various methodologies. These dimensions seek to address problems identified in the previous study, such as the need to discern between what to do and how to do it [9], giving relevance to dimension 1; the execution of non-value tasks in certain type of projects because of a rigid model [19], giving relevance to dimension 2; and the breakdown of activities without a clear structure [10], giving relevance to dimensions 3 and 4.

The catalog includes these four dimensions, allowing its users to define a methodological model that integrates best agile/traditional software development practices for projects planning. Besides, the catalog will provide the guidance to understand the implications of taking one path or another when defining the necessary activities.

4 Conclusions and Further Work

In this work, we have performed a study of agile and traditional approaches for software project management that led us to believe that it is possible to combine activities and techniques from both sides. According to this, we have made an initial proposal for the structure of a multi-model methodological catalog for software project management, which will establish activities paths guided by different dimensions that are defined within the catalog itself.

A proposal of a new methodological framework that integrates agile methodologies and traditional reference frameworks for project management could excel in providing an initial way to build the elements of a self-tailored methodology, but it also has the disadvantage of generating additional complexity, since there are a lot of possible activities to choose from, with no clear indication of how to combine them.

The multi-model catalog approach we propose allows integrating best practices from agile and traditional approaches, right through understandable steps that a company or team may follow and adjust, according to its own environment. These guided steps provide a solution to the additional complexity problem, allowing the structuring of the methodology by means of the proposed dimensions.

Our approach can help to accelerate the adoption of traditional reference frameworks by including best practices of agile methodologies. We believe this is true because, through the multi-model catalog, it is possible to facilitate the implementation of activities, eliminate activities, perform methodical replacements between activities that seek the same goal, and correlate activities of different frameworks.

When combining agile/traditional frameworks, it is important to take into account the need to carefully select the approach a project should take. It is required that the project and company characteristics are revised, so the methodology can be adapted to the project or company and not the contrary.

We are aware that the proposal is still in its early stages and there is a lot of work to do; however, we believe the structure of the catalog covers the important aspects that need to be considered when merging approaches of different nature.

As future work, we will define activities from different frameworks that will make up the catalog. We will also define guides that will provide indications for using the catalog. Also, we will define a case of study in order to use the catalog and validate it.

Our approach does not provide tools to ensure that a project is not being tailored to meet a methodology in particular. This is something we consider as further work, since we think it is important to identify when the multi-model catalog is relevant for a project context. In order to do that, we need to analyze the quality and maturity of company's processes, how such processes generate value to the company, how the company responds to requirement changes, or how much value there is in frequent deliveries to the costumer. These will help to decide whether an integrated agile/traditional framework is required or not.

References

1. Pino, F., García, F., Piattini, M.: Software process improvement in small and medium software enterprises: a systematic review. Software Qual. J. **16**(2), 237–261 (2008)
2. Institute, P.: A Guide to the Project Management Body of Knowledge (PMBOK® Guide). Project Management Institute, USA (2013)
3. Team, C.: CMMI for Development, Version 1.3. Software Engineering Institute (2010)
4. Office, I.: ISO 9001:2008 Quality management systems—Requirements, Suiza (2008)
5. Singh, J., Sharma, M., Srivastava, S., Bhusan, B.: TSP (Team Software Process). Int J Innov Res Dev **2**(5) (2013)
6. Schwaber, K.: Agile project management with Scrum. Microsoft Press, Redmond (2004)
7. Polk, R.: Agile and Kanban in coordination. In: Agile Conference, pp. 263–268 (2011)
8. Lindstrom, L., Jeffries, R.: Extreme programming and agile software development methodologies. Inf. Syst. Manag. **21**(3), 41–52 (2004)
9. Torrecilla Salinas, C., Escalona, M., Mejías, M.: A scrum-based approach to CMMI maturity level 2 in web development environments. In: Proceedings of the 14th International Conference on Information Integration and Web-based Applications & Services, New York, NY, USA, pp. 282–285 (2012)
10. Tuan, N., Thang, H.: Combining maturity with agility: lessons learnt from a case study. In: Proceedings of the Fourth Symposium on Information and Communication Technology, pp. 267–274 (2013)
11. Martin, D., García Guzman, J., Urbano, J., Amescua, A.: Modelling software development practices using reusable project pattern: a case study. J. Softw. Evol. Process **26**(3), 339–349 (March 2014)
12. Martín, D., García Guzmán, J., Urbano, J., Lloréns, J.: Patterns as objects to manage knowledge in Software development organizations. Knowl. Manag. Res. Pract. **10**(3), 252–274 (2012)
13. Biolchini, J., Mian, P., Natali, A., Travassos, G.: Systematic review in software engineering. Syst. Eng. Comput. Sci. Dept. COPPE/UFRJ Tech. Rep. ES **679**(05), 45 (2005)
14. Špunda, M.: Mixed agile/traditional project management methodology—reality or illusion? Procedia Soc. Behav. Sci. **119**, 939–948 (2014)
15. Pinheiro, P., Sampaio Machado, T., Tamanini, I.: Dealing the selection of project management through hybrid model of verbal decision analysis. Procedia Comput. Sci. **17**, 332–339 (2013)
16. Whitney, K., Daniels, C.: The root cause of failure in complex IT projects: complexity itself. Procedia Comput. Sci. **20**, 325–330 (2013)
17. Friis Sommer, A., Dukovska-Popovska, I., Steger-Jensen, K.: Barriers towards integrated product development—challenges from a holistic project management perspective. Int. J. Project Manage. **32**(6), 970–982 (2014)
18. Hornstein, H.: The integration of project management and organizational change management is now a necessity. Int. J. Project Manage. **33**(2), 291–298 (2014)
19. Monteiro, P., Borges, P., Machado, R., Ribeiro, P.: A reduced set of RUP roles to small software development teams. In: Proceedings of the International Conference on Software and System Process, pp. 190–199 (2012)
20. Buglione, L.: Light maturity models (LMM): an agile application. In: Proceedings of the 12th International Conference on Product Focused Software Development and Process Improvement, pp. 57–61 (2011)

21. Brown, A., Ambler, S., Royce, W.: Agility at scale: economic governance, measured improvement, and disciplined delivery. In: Proceedings of the 2013 International Conference on Software Engineering, pp. 873–881 (2013)
22. Ng, P.-W.: Theory based software engineering with the SEMAT kernel: preliminary investigation and experiences. In: Proceedings of the 3rd SEMAT Workshop on General Theories of Software Engineering, pp. 13–20 (2014)
23. Van Hilst, M., Fernandez, E.: A pattern system of underlying theories for process improvement. In: Proceedings of the 17th Conference on Pattern Languages of Programs (8) (2010)
24. Marino, L., Morley, J.: Process improvement in a multi-model environment builds resilient organizations. In: SEI. http://www.sei.cmu.edu/library/abstracts/news-at-sei/02feature200804.cfm. Accessed 1 Apr 2008
25. Urs, A., Heijstek, A., Kirwan, P.: A unified process improvement approach for multi-model improvement environments. In: SEI. http://www.sei.cmu.edu/library/abstracts/news-at-sei/feature1200604.cfm. Accessed 1 Apr 2006
26. Ferreira, A., Machado, R.: Software process improvement in multimodel environments. In: Fourth International Conference on Software Engineering Advances, Porto, pp. 512–517 (2009)
27. Gamma, E., Helm, R., Johnson, R., Vlissides, J.: Design Patterns: Elements of Reusable Object-Oriented Software. Addison-Wesley, USA (1994)

Situational Factors Which Have an Impact on the Successful Usage of an Agile Methodology for Software Maintenance: An Empirical Study

Lourdes Hernández, Nahum Vite, Francisco Alvarez and Alma-Rosa García

Abstract A software maintenance process implies a high difficulty degree, since it tries to keep a current version operational, concurrently with the execution of changes which once implemented must guarantee the service continuity. Therefore, the adoption of a methodology requires the accurate identification of restrictions and characteristics of software maintenance which allows anticipating risk situations and guaranteeing aim compliance. The current study examines contextual factors and its influence in a maintenance process using agile methodologies. Specifically in the maintenance of the Registry System, Assessment and on line Applicators and Auxiliaries Selection case study in the admission process to the Universidad Veracruzana. The results confirm the success of the adoption is highly associated to the identification of situational factors, to the discipline for the activity registration and control, as well as in the communication techniques and applied management.

Keywords Software maintenance · Situational factors · Agile methodology

L. Hernández (✉) · N. Vite · A.-R. García
Facultad de Estadistica E Informática, Universidad Veracruzana, Xalapa Veracruz, Mexico
e-mail: lourhernandez@uv.mx

N. Vite
e-mail: nvite@uv.mx

A.-R. García
e-mail: agarcia@uv.mx

F. Alvarez
Universidad Autónoma de Aguascalientes, Aguascalientes, AGS, Mexico
e-mail: fjalvar@correo.uaa.mx

J. Mejia et al. (eds.), *Trends and Applications in Software Engineering*, Advances in Intelligent Systems and Computing 405,
DOI 10.1007/978-3-319-26285-7_9

1 Introduction

The evolution of a software product considers the group of activities oriented to increase, adapt, correct or improve functionalities. Therefore, the usage of a software maintenance methodology which supports the control of tasks and which contributes to improve software quality is important. A software maintenance process is complex among other situations because the product must be operational and concurrently assist the requirements to include corrections or improvements [1].

It is recommended to use a methodology in software maintenance which allows minimizing risks in aim deviation. The application of agile methodologies with emphasis in frequent liberation, final user intense interaction and iterative development is a useful option for projects with time restrictions and reduced developmental teams [2].

In a software maintenance project converging factors of a different nature from human to technology. Determining the most appropriate strategy to perform depends on the type of software maintenance and the execution context. A successful adoption strategy of a process depends on specific characteristics of the maintenance project that is under attention. Among such characteristics, the personnel experience on business control, technical competence on development tools to be applied or requirement stability can be mentioned.

There are frameworks of reference [1, 3, 4] which put forward a group of factors that are related to software processes, oriented to facilitate the quick identification of restrictions and characteristics of a specific project. The current study takes Clarke's et al. [4] proposed frameworks as the main point; as well as the factors related to sustained adoption of an agile methodology suggested by Senapathi et al. [2] and it also selects several applicable factors to a case study in order to validate its dependence with the project's compliance on time, by the application of agile methodologies for the control and compliance of the requested maintenance process.

The information presented within this article is structured in the following sections: related research work, design and compliance of the case study, results analysis and conclusions.

2 Related Research Work

The maintenance of a software product implies an important resource revision and it also has an impact on the business objectives of the institution or company which it is used in. A software maintenance process includes the completion of the stages [2] shown on Fig. 1b.

In a software project different kinds of factors converge, from human to technological resources, so the precise software maintenance identification and its realization context nature is necessary in order to determine the most appropriate strategies for a successful compliance [5–7]. The revision of related research work

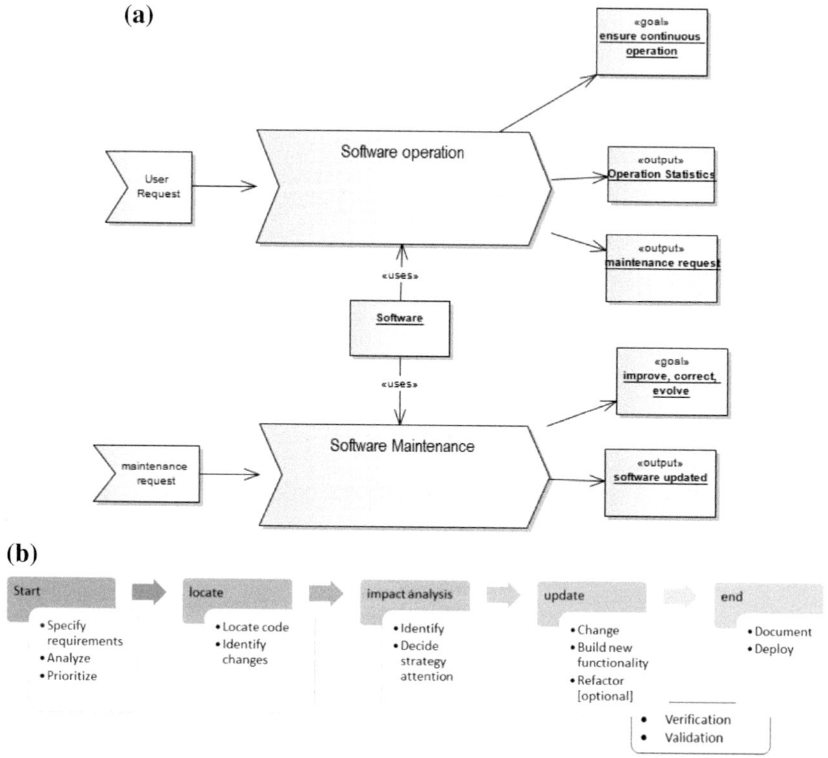

Fig. 1 **a** The software maintenance process is directly related with the operational process itself, increasing its complexity in the incorporation of updates, **b** Stages of a maintenance process

was guided by the question: Which are the factors to support the decision making related to the implementation of an agile methodology in a software maintenance process.

The search range considered publications from January 2010 to January 2015. The consulted databases were: IEEE Xplore, ACM digital Library, Science Direct (Elsiever), Web of science. The search equation was structured as follows: ("software maintenance process" or "software evolution process") and ("migration" or "adoption" or "introduction") and ("strategy" or "feasibility" or "barriers" or "success factors"). From these initial results, those non-oriented to an evaluation or identification of success factors were excluded from the methodology application. Eleven of the research work which included a systematic revision, improvement processes and case of study were analyzed in detail.

Rajlich analyzes the importance of a software maintenance process, pointing that before a dynamic context in the technological and business process aspects, such activity takes particular relevance [8]. He reflects about the applied techniques, especially those of the agile type and he also presents a reflection about the Software Engineering teaching process.

Clarke et al. proposed an analysis through the inherent factors of situational characteristics associated to a specific software developmental or maintenance scenario [1, 4]. For instance, the following can be mentioned: the nature of the system, the size of the hardware, the volatility of the requirements and personal experience. Clarke's et al. work is supported on an analysis of 22 elements of research work related to seven specific domains, which results is an initial framework of 8 factors and 44 characteristics [1, 3].

In Sulayman's et al. work [7] a qualitative study is described. It determines the success key factors in the implementation of improvement strategies in small companies (10 to 50 employees) devoted to web development. Some of the identified factors include: automatic tool usage to improve processes, metric definition and usage, improvement process activity definition and implementation of iterative strategies, project monitoring through revisions, standards implementation (e.g. CMMI [9] and IDEAL [10]), final user and development team satisfaction, feedback mechanisms, management support, process priority, process documentation, improvement process alignment with the business objectives as well as aim and benefit achievement of the improvement process. On the other hand, Pedreira et al. [6] analyze the impact of rule application on games, for example, points and insignia in software engineering processes.

In reference to factors associated to agile methodologies application, Senapathi et al. [2], introduce an analysis of factors associated to the adoption and sustained usage of agile methodologies.

The findings on the revision of selected work make the following situations evident:

- The strategies have a relevant emphasis towards a software development process, and less on maintenance processes characteristics or software evolution.
- The proposed frameworks of reference on the analysis of situational factors which have influence on a software developmental or maintenance process, have recently appeared, and therefore validation is required.

The problem presented in this work validates the compliance strategy of a software maintenance process, specific situational factors, and the project success degree to support, from the results, the decision making towards management and operation processes.

3 Methodology

The process to perform the exploratory study current in this research work is introduced in Fig. 2. Using the standard notation BPMN (Business Process Model and Notation) [11] describing the activities to perform, the relation between them and the expected products.

The current study introduces an evaluation of the result of the aspects of software maintenance process control applying the MANTEMA [12] based Software

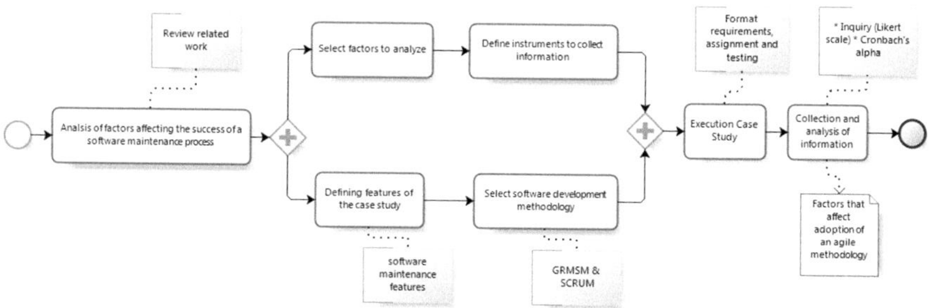

Fig. 2 Methodology for the implementation of the current study using standard BPMN. The task sequence and the main products, derived from each of them is indicated

Maintenance Reduced Guide methodology (GRMSM, for its acronym in Spanish; MSMRG hereinafter) in the development of the second version of Application and on Line Auxiliaries Evaluation and Selection Registration System (APLIAU-UV V 2.0 for its acronym in Spanish,), which participate in the entrance examination for undergraduate and technical college levels of the University of Veracruz, Mexico. The main objective of the study is to determine the continuity of the usage of MSMRG and its contribution in the objective business. A group of functional and non-functional requirements frame the current work, and the MSMRG implementation setting the proper scenario to analyze its application.

The verification instruments of time and range variables are characteristic of the MSMRG and SCRUM methodologies, while the instruments to measure qualitative situational factors were selected by its nature, such as design and implementation of a survey and statistic techniques to analyze the results.

4 Case of Study Design

The case of study design considers the definition proposed by Runesco et al. [10], "The case of study in software engineering is an empirical proposal which uses multiple evidence sources to research an instance (or a reduced number of instances) of a contemporary software engineering phenomenon and its relation in a real implementation context, especially when the limits between the phenomenon and the context cannot be clearly specified". The current case of study is conceptualized as a specific proposal of software product maintenance used by an institution of higher education and supported by a reduced group of people having an impact on important objectives of the educational business. The most important elements of the present case are summarized as follows.

Essential Reason: to establish a strategy to maximize the probability of success in the APLIAUX.UV software maintenance.

Goal: to determine the implementation continuity of the MSMRG methodology.

Case: improvement of the trainers' management module of the APLIAUX-UV system.

Study Unit: Coordination of school admission of the University of Veracruz, group responsible of software development and maintenance. APLIAUX-UV system.

Related research work: analysis of research work about factors which have an impact on the implementation of methodologies in software development or maintenance.

Research question: Which are the factors to consider for the determination of continuity in the implementation of an agile methodology in a software maintenance process.

Hypothesis: a group of contextual and technical factors determine the success in the usage of a software maintenance methodology, subject to a dynamic environment which will be assessed before deciding the implementation of a specific methodology.

Concept and measurement definition: definition of new functionalities through use cases, function points effort measurement. Effort days estimated time unit, with a upper limit of 6 months. Eleven people participated with the roles of applicant, final user, maintenance responsible, developer and test engineers. Contextual impact factors definition.

Methods and data extraction: the information was extracted from four models: UML (use cases, collaboration, sequence, class). Six formats ruled by MSMRG. The Sprint Backlog, Retrospective and Product backlog from SCRUM. The evaluation of contextual or situational factors was implemented through a survey using a Likert scale. The validity of the survey was verified applying the Alpha Cronbach technique.

4.1 Software Maintenance Project Context

The group of characteristics which dominated during the case of study implementation process is listed as follows.

- An exploratory program of the possible technologies was not required since the same technology of the previous version of APLAIUX-UV was applied.
- There was a minimum gap between the ability shown by the programmers and the expected one, assisted by the maintenance for the control of change incorporation responsible.
- Human resources participant rotation was inexistent in the estimated time period for the maintenance implementation.
- The selected technological tools showed a maturity and stability degree convenient for the project.
- The team training on agile methodologies required gradual training as every phase of the process went by.

4.2 Functional and Non-functional Requirements

The specific maintenance request consisted on improving and adapting the following functionalities: (a) evaluate the participants with a random assessment on line, in order to know the learned abilities during the training course, (b) to implement a pre selection process and arrangement of the participants for their training, distributing them evenly in the different sites of participation.

For non-functional requirements, the following were considered: an intuitive user interface, a standard security level (SQL injection or Cross-site Scripting attacks) [12] and support for easy modification of its modules. The appointed norm of the Coordination of University Transparency, information access, and personal information protection was applied, following the 581 Law for the protection of the Personal Information for the State of Veracruz, Mexico.

4.3 Software Development Applied Methodologies

Due to the characteristics of the case of the study and institutional guidelines, the MSMRG was selected, proposed by Viveros [12]. It includes the following stages: Problem Definition, Analysis, Design-implementation, Test, Liberation and Final Assessment. Each stage had a format for task control and management (Fig. 3).

According to Shuterland [13], SCRUM is a working framework which allows managing the development of a software product. For the case of study, the Sprint Backlog, Retrospective and Product backlog were used. Nine sprints were

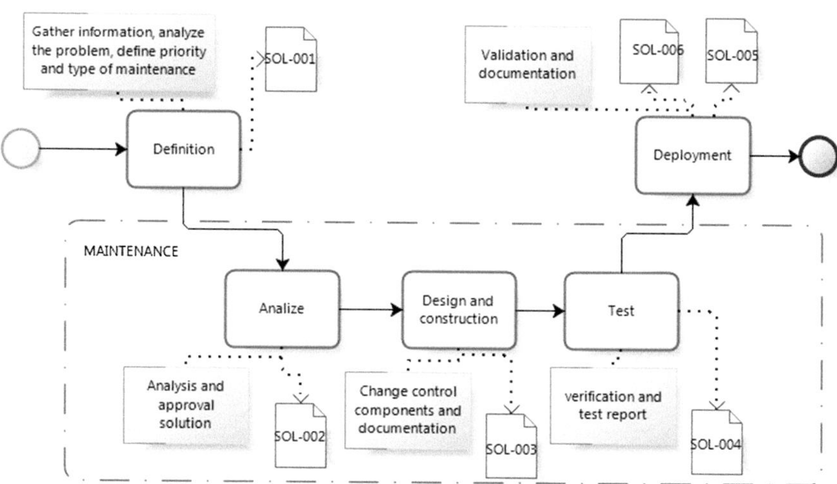

Fig. 3 Software maintenance reduced guide methodology (GRMSM. The task sequence and the main artifacts is indicated

Sprint 8. Integrate modules to production system- 6 February to 24 February 2014
Objective: To integrate the preselection view, accommodation and random test UV-APLIAUX 2.0. and generate the code validation

ID	Use Case / task	sprint days / effort to make																				Category
		0	1	2	3	4	5	6	7	8	9	10	11	12	13	14	15	16	17	18	19	
		40	39	39	39	39	39	39	39	39	39	39	34	34	34	25	20	12	6	1	0	
2	**1.-Review random online,** **2.- Preset and accommodation of the participants**																					
	Integrate (view and functionality) preselection and accommodation to APLIAUX-UV2.0 module and Add button in the administrator control panel	8	8	8	8	8	8	8	8	8	8	8	7	7	7	5	4	4	2	1	0	Technical task
	Integrate module (view and functionality) of random tests online at APLIAUX-UV2.0	8	8	8	8	8	8	8	8	8	8	8	7	7	7	5	4	2	1	0	0	Technical task
	Investigate and implement date validation in grocey crud	8	7	7	7	7	7	7	7	7	7	7	6	6	6	5	4	2	1	0	0	Knowledge acquisition
	Investigate and implement validation in 24-hour format	8	8	8	8	8	8	8	8	8	8	8	7	7	7	5	4	2	1	0	0	Knowledge acquisition
	Tests and changes	8	8	8	8	8	8	8	8	8	8	8	7	7	7	5	4	2	1	0	0	Technical task

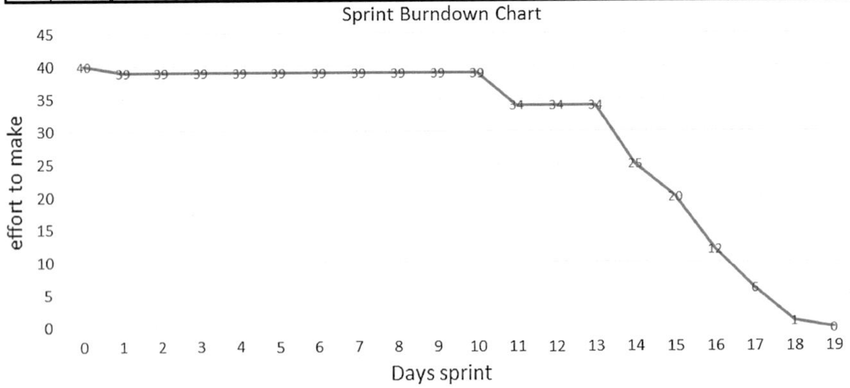

Fig. 4 Scrum methodology. Example of activities and sprint burndown chart

generated for three usage cases (See Fig. 4). The duration of each sprint was planned from 15 to 20 days ideally and for each task some previously established difficulty points were assigned. Product backlog was used to visualize the aims of each sprint and the difficulty points were taken from Sprint backlog to complete each of the tasks. Finally, retrospectives were held for each sprint, in order to continuously improve the deliveries and quality of the product. Reunions with the final users were held in order to validate the requirements and to identify adjustments as early as possible.

Analysis of sprint burndown chart reflects the experience of the team in the scope setting to achieve the goals. The graph shows a late progress in the first half of the sprint where the team met some issues to ensure the achievement of the goal. As a result of the retrospective the team improved in the execution of the next sprint and reconsider their ability to complete additional work.

4.4 Analyzed Situational Factors

From the analysis of some related work, the proposals of Senapathi [2] and Clarke [4] were considered to select a set of factors that affect the determination of relevance for the continuity of application of an agile methodology for software development. From both proposals, matches with the presented case study were

Table 2 Selected factors and indicators for perception assessment given by the members of the maintenance team

Factors	Indicators													
	A	B	C	D	E	F	G	H	I	J	K	L	M	N
F1. Management support	X													
F2. Attitude		X	X	X	X									
F3. Motivation						X	X							
F4. Team composition								X	X	X		X	X	
F5. Training											X			X
F6. Agile mind												X	X	
F7. Engineering practice														X

The columns that correspond to the indicators are identified by *A* Communication, *B* Honesty, *C* Collaborative attitude, *D* Learning willingness, *E* Responsibility, *F* Recognition, *G* Growth expectations, *H* Technical abilities, *I* Decision making, *J* business control knowledge, *K* Training, *L* Team commitment, *M* Institutional belonging sense, *N* Tool usage

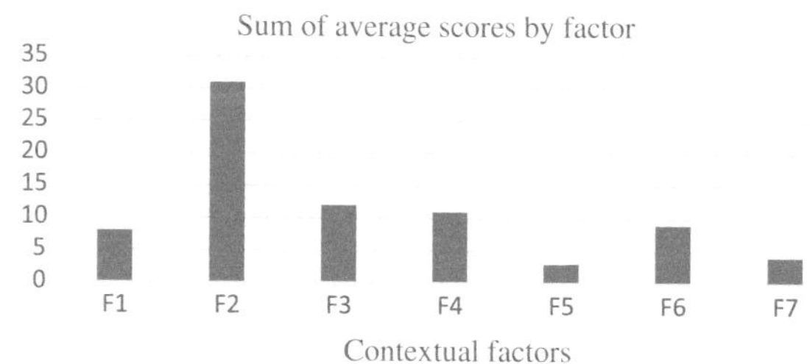

Fig. 5 Grouped grades by assigned factors by the maintenance team comprised by eleven people

identified and it was determined to carry out a survey among the participating members obtain their perception. Thereon seven of these factors were associated with indicators as shown in Table 2 and Fig. 5. The survey was composed of twenty questions using a rating scale from 1 to 5, where 5 represented the highest impact or relevancy.

5 Results Analysis

The implementation of MSMRG and SCRUM supported the communication between the project leader (responsible) and the maintenance applicant (user), specifically in the requirement definition, the recording of the request and liberation dates of each of the devices and the recording of the delivery dates of the working software. It supplied the evidence registry and, consequently, the convenient failure detection.

The established guidelines allowed maintaining a collaborative and supportive environment among team members, favoring a constant rhythm of productivity and responsibility, especially within the short delivery cycles. The recognition to the daily effort by each participant was shown in the daily pursuit and in the retrospective meetings. The involved personnel in the case of study kept a balance between technical competence and detailed knowledge of thc business control and a series of relevant pieces of evidence according to the process key elements.

Due to the obtained results, the team was willing to learn the agile techniques and to perform tasks as indicated by the maintenance responsible. The final assessment of situational factors confirmed their relevance from the perspective from all the participants.

The results presented a high consideration of factors impact attitude motivation and composition of the team. The applied methodologies encourage collaborative work and communication. The project manager role confirmed the impact of the factors supporting the management, training, agile mindset and practices of software engineering. The managers promptly detected deviations from the objectives and verified compliance responsibilities through tools mapping the time and effort invested. This situation facilitated the implementation of preventive and corrective actions as proper training and delimitation of functional scope.

6 Conclusions and Future Work

When closing the analysis, the importance of having methodological tools is clear, as they allow an adequate management of software maintenance, leading efforts to reach the foreseen targets and thus the fulfillment of the business objectives. The contextual factors analysis allowed a successful planning and performance considering particularly to even the experience and expertise levels of the team, a recognition of the performed work, growth expectations, effective communication and risky situations early detection.

Nevertheless, it is important to point that the success in the current case study must be taken with reserve when modifying the compliance context. In such case the proposed criteria must be taken as referent to minimize the impact and to guarantee proper performance. The frameworks taken as reference highly facilitated the identification of characteristics and restrictions associated to the project.

For the future work lines, the automation of MSMRG flow can be mentioned, considering the six used devices and the analysis of key indicators, associated to processes to optimize, along with the frameworks of reference, the convenient detection of objective deviations.

References

1. April, A., Abran, A.: A software maintenance maturity model (S3 M): measurement practices at maturity levels 3 and 4. Electron. Notes Theoret. Comput. Sci. **233**(C), 73–87 (2009)
2. Senapathi, M., Srinivasan, A.: Sustained agile usage: a systematic literature review. In: 17th International Conference on Evaluation
3. Clarke, P., O'Connor, R.V.: The influence of SPI on business success in software SMEs: an empirical study. J. Syst. Softw. **85**(10), 2356–2367 (2012)
4. Clarke, P., O'Connor, R.V.: The situational factors that affect the software development process: towards a comprehensive reference framework. Inf. Softw. Technol. **54**(5), 433–447 (2012)
5. Jabangwe, R., Börstler, J., Šmite, D., Wohlin, C.: Empirical evidence on the link between object-oriented measures and external quality attributes: a systematic literature review. Empirical Softw. Eng. 1–54 (2014)
6. Pedreira, O., García, F., Brisaboa, N., Piattini, M.: Gamification in software engineering—A systematic mapping. Inf. Softw. Technol. **57**, 157–168 (2015)
7. Sulayman, M., Urquhart, C., Mendes, E., Seidel, S.: Software process improvement success factors for small and medium Web companies: A qualitative study. Inf. Softw. Technol. **54**(5), 479–500 (2012)
8. Rajlich, V.: Software evolution and maintenance. Proc. Future Softw. Eng. FOSE **2014**, 133–144 (2014)
9. Senapathi, M., Srinivasan, A.: CMMI Development, "CMMI® for Development, Version 1.3 CMMI-DEV, V1.3," no. November, 2010. Sustained agile usage: a systematic literature review. In: 17th International Conference on Evaluation and Assessment in Software Engineering (EASE'13), pp. 119–124 (2013)
10. Mcfeeley, B., Sulayman, M., Urquhart, C., Mendes, E., Seidel, S.: "IDEAL: a user's guide for software process improvement," no. February, 199611. Software process improvement success factors for small and medium Web companies: a qualitative study. Inf. Softw. Technol. **54**(5), 479–500 (2012)
11. OMG: Business Process Model and Notation (BPMN) (2011)
12. Viveros, A.: Guia Metodologica para el Mantenimiento de Software. Universidad Veracruzana, Facultad de Estadistica e Informática (2010)
13. Sutherland, J., Schwaber, K.: The definitive guide to scrum: the rules of the game (2013)

Definition and Implementation of the Enterprise Business Layer Through a Business Reference Model, Using the Architecture Development Method ADM-TOGAF

Armando Cabrera, Marco Abad, Danilo Jaramillo, Jefferson Gómez and José Carrillo Verdum

Abstract Definition and implementation of the Enterprise Business Layer, is focused on strengthening the business architecture and IT governance. This work allows the company to define, and to be clear about the key elements of the business, also allows to establish the business baseline and the target business state. This work uses ADM-TOGAF v9. Based on this, the iteration of architectural capability of ADM has been tackled (Preliminary Phase and Architecture Vision), that has allowed the creation and evolution of the architectural capability required to start the exercise of enterprise architecture, this includes the establishment of an approach, principles, scope, vision and architectural governance. The architecture capability iteration serve as input for the first phase of the development iteration (Business Architecture), in which it has been performed the AS-IS and TO-BE analysis "as reported by Giraldo et al. (Propuesta de arquitectura empresarial proyecto banco los alpes)" [1], which enables to establish the key points that should a company take to improve their current business scheme.

A. Cabrera (✉) · M. Abad · D. Jaramillo · J. Gómez
Universidad Técncia Particular de Loja, 1101608 Loja, Ecuador
e-mail: aacabrera@utpl.edu.ec

M. Abad
e-mail: mpabad@utpl.edu.ec

D. Jaramillo
e-mail: djaramillo@utpl.edu.ec

J. Gómez
e-mail: jfgomez@utpl.edu.ec

J.C. Verdum
Universidad Politécnica de Madrid, 28040 Madrid, Spain
e-mail: jcarrillo@fi.upm.es

J. Mejia et al. (eds.), *Trends and Applications in Software Engineering*,
Advances in Intelligent Systems and Computing 405,
DOI 10.1007/978-3-319-26285-7_10

Keywords Enterprise architecture · Business reference model · Business strategy

1 Introduction

Enterprise Architecture (EA) currently plays an important role in the structure of organizations [2], due to the constant evolution of society, the advance of competitors and market requirements, the companies must be prepared to endure the changes and be at the forefront of technology, the level of the competition and the demands of customers [3].

The current concept of Enterprise Architecture (EA) implies a form of view the organization in an integrated and relational way, considering each and every one of the elements that compose it [4]. This leads to establish a business vision for business transformation that helps to align technology with the needs of the organization from a strategic perspective and customer oriented.

The organization's ability to manage business transformation is highly necessary to stay competitive, this involves fundamental and complex organizational changes, not only within companies, but also through their value chain. This can also change the relationship between a company and its broader economic and social environment. The transformation process is complex, time consuming and always occurs within an ecosystem that is organization's core disciplines and its environment (i.e. customers, competitors, government regulators and vinvestors). This process only can be done handling an EA framework.

MALCA Company was used as a case of study, which is an industry dedicated to the production of sugar cane derivatives, and food marketing. Its main offices (administrative, financial), field and factory are located in Loja—Ecuador [5]. Currently the capacity in terms of infrastructure of the company is limited, and the business processes are not aligned with their strategic business goals.

Therefore, it is necessary to evaluate the architectural capacity of the enterprise and project a vision to bring the technology investment, requirements, applications, process changes, business modeling, market factors, resources (human, technical) all available in each area to meet comprehensively and coordinately the goals of the company.

2 Related Work

Lots of literature has been written in order to guide managers and their companies through a process of change. From a practical point of view, understanding the key concepts are well established, but their integration and implementation within programs of transformation is less known. From an academic point of view, the theoretical work on the transformation has coalesced around a set of architectural frames, but not taken into consideration practical aspects of linking elements of a coherent and effective way, especially when it involves coping to complex transformations.

3 Problem

MALCA currently lacks an organizational model that integrates coordinately the IT with their business vision, this has led to the company lacks of: alignment between resources and standardized development policies, of agility in making business decisions and performance information of their areas of production, administrative, financial and IT. For this work it is proposed to use the architectural development method TOGAF ADM, which will allow to:

- Implement a system of controls on the creation and monitoring for all components of the architectural exercise activities to ensure the successful introduction, implementation and evolution of the architectures within the organization [6].
- Implement a system to ensure compliance with internal and external standards and regulatory obligations.
- Establish, define and develop processes to support effective management of corporate activities within agreed parameters.
- Develop practices that ensure accountability to stakeholders clearly identified.
- Liaising between processes, resources and information with business objectives and strategies.
- Integrate and institutionalize the best architectural practices
- Align with IT management frameworks such as COBIT (planning and organizing, acquiring and implementing, delivering and supporting, and monitoring the IT performance) [7]
- Maximize the information, infrastructure and assets of hardware/software.
- Protect digital assets underlying the company.
- Support (internal and external) regulations and better practices such as auditing, security, responsibility and accountability.
- Promote risk management [7, 8].

3.1 TOGAF

TOGAF is an architectural framework, which provides a focus for the design, planning, implementation and governance of an EA. The framework is oriented toward the four levels or business domains: Business, Data, Applications and Technology [9]. Its iterative model based on good practices allows the acceptance, use, and maintenance of enterprise architectures using the ADM as a primary method of development [9].

Table 1 TOGAF structure, adapted from [6]

Section	Description
Part I: Introduction	Provides a high-level introduction to the key concepts of Enterprise Architecture for key stakeholders and in particular to the TOGAF approach. It contains definitions of terms used throughout TOGAF and release notes detailing the changes between this version and the previous version of TOGAF
Part II: Architectural development method	Architectural development method is the core of TOGAF. Describes the development method of TOGAF architecture—is a gradual approach to the development of an enterprise architecture
Part III: ADM guidelines and techniques	Contains a collection of guides and techniques available for the implementation of ADM
Part IV: Content framework	Describes the architectural framework of TOGAF content, including a structured meta model for architectural artifacts, the use of Architectural Building Block (ABB) and a description of typical architecture deliverables
Part V: Enterprise continuum and tools	Addresses the appropriate taxonomies and tools for sorting and storing the results of the architectural activity within an enterprise
Part VI: TOGAF reference models	Provides two architectural models of reference: technical reference model and TOGAF reference model for integrated information infrastructure
Part VII: Capability framework	Capacity analyzes the organization, processes, capabilities, functions and responsibilities required to establish and operate an architectural function

TOGAF reflects the structure and content of the Architectural capacity within a company, Table 1 describes each of this components.

ADM suggests a series of artifacts and deliverables that has been adapted to the reality of business, they are the product of architectural exercise and aim to specify contractually and describe the key aspects of the project. Figure 1 shows the phases and iterations of ADM.

In the developed architectural exercise the capability iteration was applied, it considers the Preliminary and Architectural Vision phases. Throughout the development of these phases validation of results were made with the main stakeholders of the business case, it has allowed to consider and validate the project's scope.

4 Activities and Results

The activities and results have been obtained by applying the guidelines of ADM, and using TOGAF templates which have been adapted to the reality of the case under study.

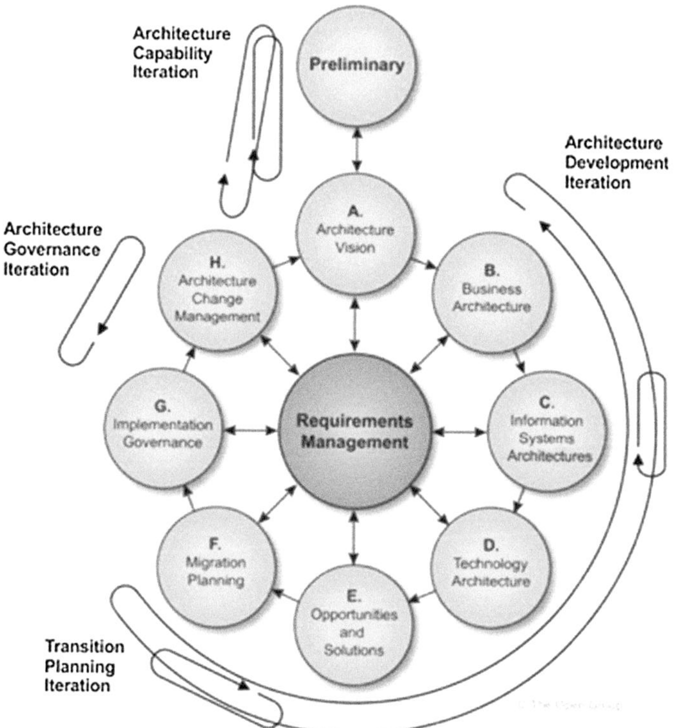

Fig. 1 Phases of ADM, adapted from [6]

4.1 Activities

The activities within the study case to get the results are described below:

Preliminary phase: In this phase the activities were (i) Define the scope of the affected business organizational units; (ii) Check the governance and support frameworks; (iii) Define and establish the organization and equipment of the enterprise architecture; (iv) Identify and define the architectural principles; (v) Tailor TOGAF; (vi) Implement architectural tools;

Architectural vision phase: In this phase the activities were (i) Set the architectural Project; (ii) Identify the stakeholders concerns and business requirements [10, 11]; (iii) Confirm and develop the business goals, business drivers and constraints [12]; (iv) Evaluate the enterprise capability; (v) Evaluate the willingness of the company to business transformation; (vi) Define the scope; (vii) Develop architectural vision; (viii) Define the value propositions of the target architecture; (ix) Identify the risks of business transformation and mitigation activities (developed conceptual level); (x) Develop the statement of architectural work.

Business architecture phase: In this phase the activities were (i) Select reference models, points of view and tools; (ii) Develop the description of the business architecture baseline; (iii) Develop the description of the target business architecture; (iv) Conduct gap analysis [13]; (v) Define the components of the architectural roadmap candidates; (vii) Solve architectural landscape impacts; (viii) Conduct formal review with stakeholders; (ix) Finish the Business Architecture; (x) Create the document Architectural Definition.

4.2 Results

The development of the activities described above in the study case has yielded the following results:

4.2.1 Preliminary Phase

Evaluation of architectural capability, the main issues of the study case were identified:

4.2.2 Insufficient Level of Architectural Definition

- There is no mandate for EA.
- The EA is seen as an isolated IT activity.
- The EA (standards, references, drawings) is performed in silos without having a comprehensive view of the company.
- The EA generates no impact either on business or results of IT.
- Architectural Governance is nonexistent, IT guiding principles, reference architectures and architectural schemes are not considered
- Little or no exchange of business and technology platforms (except perhaps the infrastructure).

4.2.3 Ineffective or no Architectural Organization

- The skills and knowledge architectural are absent, underdeveloped or unused.
- Informal or undefined processes for the management or delivery of architectural projects.
- There is not an area within IT architecture focused on enterprise management.
- The EA is activity that has little or no participation in the value chain.
- Incentives focused on the performance of functional units and business, any at enterprise level.

4.2.4 Lack of Alignment Between Business Activities and IT

- Culture of "firefighters" instead of proper planning.
- Integrated plans (strategic plan 2008, Ishikawa´s plan) do not have support of business stakeholders and IT.
- The role of the EA lacks clarity with regard to operational and business stakeholders, and its decision-making process.
- Pressure to implement short-term requirements without long-term vision.
- The business areas implement their own technologies (SIGTH, Integrated System of Human Resource Management), without taking into account business considerations.
- Proliferation of applications and redundant processes.

It was identified, mapped out and adapted an architectural framework [14] (Fig. 2 shows the basic components of the architect's office), which has created the organizational business architecture through the architectural office. The establishment of this office in the case study allowed:

- Create awareness and architectural culture in major business stakeholders
- Manage the entire architectural exercise
- Focus architecture in different value propositions according to business work environments: Strategic Alignment Transformation Plans and Project and Portfolio Management
- Transform the business structure from an IT management scheme into a comprehensive management scheme and transformation at organizational, process and IT.
- Link the gaps between business strategy and IT
- Generate benefits: Project Management, Business Management and Governance.
- Adding value to business management

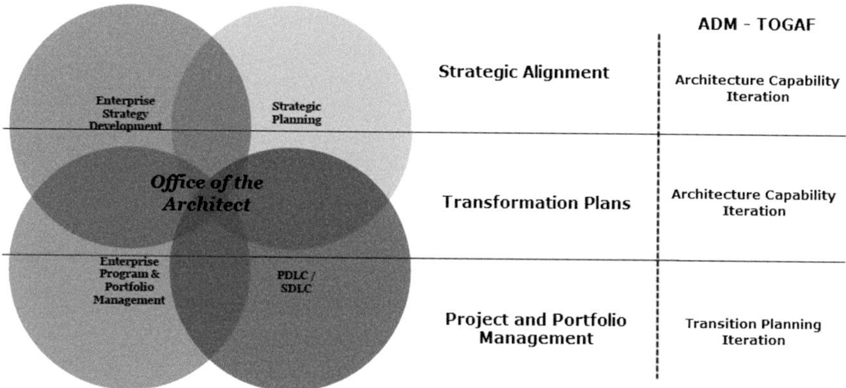

Fig. 2 Office of the achitect, adapted from [14]

- Evaluate the architectural capacity through maturity models.
- A set of architectural principles were defined for domains/subdomains based on the requirements of the company:
- Principles of Business: Business Planning, Common Vocabulary, Simple and flexible technological independence, centered on the client.
- Principles of Data and Information: Formally defined, alignment with business needs, Clarity and Consistency, integrity, accessibility and availability.
- Principles of Applications: Traceability, Flexibility, Modularity, Buy versus Build, consolidation, interoperability, reusability, compatibility, Updates, Compliance, Security.
- Technology Principles: Property, Technology Business Model Integration Level Approach Quality Metrics Maintenance infrastructure, rationalization and Platform Products, Product Selection, Product Portfolio, Infrastructure, Security/Privacy, Design, Strength and Resilience.

Organizational Model: organizations (departments, sections, sub-sections) of the company affected by the architectural work were described.

Adapted frame Architectural Reference: It allowed TOGAF to be used effectively with other frameworks such as CMMI process, project management like PRINCE2, PMI (PMBOK) and IT governance also some operational like COBIT and ITIL.

Application of architectural work: business imperatives behind the project were described, thereby driving the requirements and performance metrics for architectural work.

4.2.5 Architectural Vision Phase

Stakeholder's matrix: the main involved ones were recognized and defined in the architectural exercise; also it allowed to have a view of high-level issues and concerns that are currently have main stakeholders in the case under study [10]. Additionally allocation of responsibilities were performed (RACI) which set, to each of the components of reach specific role [15].

Architectural Vision: the expected result to be achieved in the architectural exercise was defined; it gives communications supports of the project by providing an initial version on a level of an executive summary of the architectural definition version, the same that was socialized to each of the stakeholders of the business.

The settings and models proposed for the case study process, is shown through a value chain (Fig. 2) as a part of the architectural vision (Fig. 3).

4.2.6 Business Architecture Phase

Business Model Reference: It provided the guidelines that the company used for the development of the assets of EA [17].

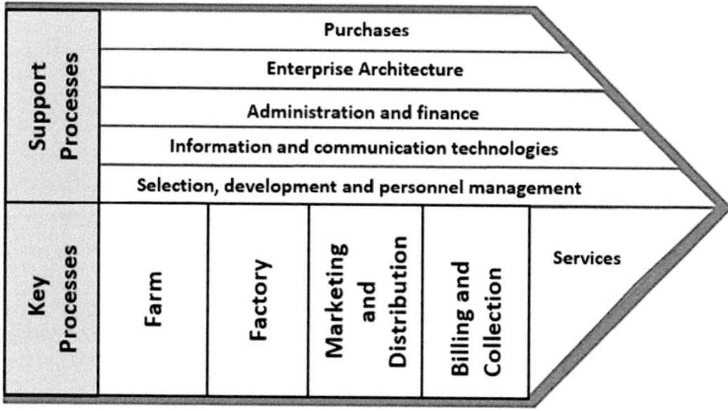

Fig. 3 Value chain. Adapted from [16]

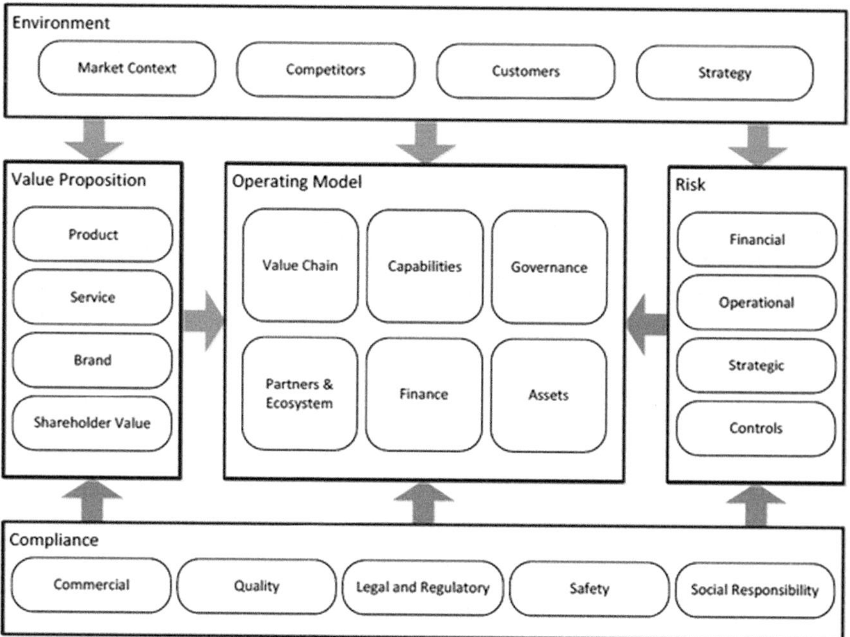

Fig. 4 Business model. Adapted from [17]

Figure 4 proposes a reference model for managing the business, this model was evaluated by the major business stakeholders for its implementation.

Table 2 describes the activities related to each of the components of the reference model and how they were implemented within business.

Table 2 Description of the outlook business model

Perspective	Description
Environment	Addresses the context in which the company will operate. External factors affecting performance were identified: competition, regulation and customers of the company, besides it was described the overall strategy to be implemented by the company for its market positioning. This perspective describes why the company has the motivation to follow a certain course of action
Value proposition	Describes the offer made by the company in terms of products, services, brand and value to its shareholders. This perspective specify the impact that the company wants to generate and how it should add value to key stakeholders
Operating model	Describes the resources available to the company, the same should be implemented to generate the value proposition. This perspective describes how the company will be able to meet their value propositions
Risks	Identifies the main risks existing around how the company delivers its value proposition uncertainties. This perspective describes the threats that face the company from "inside out" (INSIDE-OUT)
Compliance	Represents the set of criteria that the company must meet in order to ensure its value proposition will be delivered with acceptable business practice. This perspective described the restrictions that prevent the company act in negative, destructive or improperly way and the corresponding opportunities that can be exploited from a discrete position of compliance

5 Conclusions

An exercise of business transformation depends on the vision that IT leaders and managers provide to an organizational transformation project, where the establishment of the architect's office is fundamental to address business problems through an architectural framework. The company also must be managed on the basis of a reference model which is suited to its current state, this model should be launched in a long-term projection, and it will improve the business structure in which the company acts. Finally the adoption of the process reference model can be executed at any time.

References

1. Giraldo, L., Balaguera, N., Rodríguez, P.: Propuesta de arquitectura empresarial proyecto banco los alpes. Bogotá, Colombia: ECOS, 14 de Marzo del (2010)
2. Cabrera, A., Quezada, P.: Information Technology Management Guide, Loja-Ecuador (2013)
3. Ross, J., Peter, W., Robertson, D.: Enterprise architecture as strategy: creating a foundation for business execution. Cambridge, Massachusetts: Hardcover, Agosto 1 del (2006). 8398-HBK-ENG
4. ISO 42010:2007: Systems and software engineering—recommended practice for architectural description of software-intensive systems, Edition 1. Ginebra, Suiza (2011)

5. MALCA CÍA. LTDA. http://www.malca.ec/informes/index.php/corporativo/resena-historica. [Cited: 24-02-2014.] http://www.malca.ec. Accessed Oct 2014
6. THE OPEN GROUP: TOGAF® Version 9.1. USA. 2011. 978-90-8753-679-4
7. ISACA: A business framework for the governance and management of enterprise IT, COBIT 5. Rolling Meadows, Estados Unidos (2012). ISBN 978-1-60420-282-3
8. Stefan, K., Andrea, K.: Introduction to ITIL® Versión 3 and ITIL® Process Map V3. Hörnle, Alemania, Pfronten (2010). D-87459
9. Infantil, L.: TOGAF® Versión 9.1—Pocket Guide. s.l.: Van Haren Publishing, Zaltbommel (2013). ISBN: 978 90 8753 813 2
10. PMI: A Guide to the project management body of knowledge (PMBOK® Guide) fourth edition. Philadelphia, EEUU: s.n. (2008). ANSI/PMI 99-001-2008
11. Gottesdiener, E: The software requirements memory Jogger™, 1st edn. Salem, Massachusetts, EEUU, Goal Q P C Inc, Noviembre 30 del 2005. ISBN 1-57681-060-7
12. International Institute of Business Analysis. A Guide to the Business Analysis Body of Knowledge® (BABOK® Guide) V2.0. Toronto, Canadá: s.n. (2009). ISBN-13: 978-0-9811292-2-8
13. Analysis, Gap: Methodology, tool and first application. Moritz. Darmstadt, Alemania. Septiembre del, Gomm (2009)
14. Beshilas, W.: Creating a Sustainable Architecture Organization. Chicago, EEUU: Saturn Conference, 7 Mayo del (2014). 847-274-3071
15. Logarini, C.: RACI Matrix, una herramienta para organizar tareas en la empresa. Buenos Aires, Argentina (2009)
16. Porter, M.: Competitive Advantage Creating and Sustaining Superior Performance. New York, EEUU: The Free Press, 1 de Junio de (1998). 0684841460
17. The Open Group, BRM. World-Class EA: Business Reference Model. San Francisco (2014). Document No.: W146

Part II
Knowledge Management

Towards the Creation of a Semantic Repository of iStar-Based Context Models

Karina Abad, Juan Pablo Carvallo, Mauricio Espinoza and Victor Saquicela

Abstract The System Architecture definition of the majority of organizations requires a deep understanding about its environment and organizational structure; nevertheless, the construction of such models is not an easy task because of the gap of knowledge and communication between the technical and administrative staff. The DHARMA method, which makes intensive use of the *i** notation, propose a solution to that problem, supporting the Context Models construction. This paper presents an approach for annotating the *i** models with semantic technologies, to support the search and generation of Context Models based on the knowledge defined in ontological models, allowing the creation of a Semantic Repository of Context Models, which will be used to discover relations between different *i** models and extract patterns from them.

Keywords Context models · Context actors · Strategic dependency · Ontology · Semantic repository · Linked data

1 Introduction

Modern enterprises rely on Information Systems (IS) designed to manage the increasing complexity of interactions with its environment and operation. The Enterprise Architecture (EA) [1] is an approach increasingly accepted, that com-

K. Abad (✉) · J.P. Carvallo · M. Espinoza · V. Saquicela
Departamento de Ciencias de la Computación, Universidad de Cuenca, Cuenca, Ecuador
e-mail: karina.abadr@ucuenca.edu.ec

J.P. Carvallo
e-mail: pablo.carvallo@ucuenca.edu.ec

M. Espinoza
e-mail: mauricio.espinoza@ucuenca.edu.ec

V. Saquicela
e-mail: victor.saquicela@ucuenca.edu.ec

J. Mejia et al. (eds.), *Trends and Applications in Software Engineering*,
Advances in Intelligent Systems and Computing 405,
DOI 10.1007/978-3-319-26285-7_11

prises different levels of architectural design, which starting from the business strategy allows to identify IS Architecture. Early phases of EA are usually oriented to model the enterprise context, aiming to understand the purpose of enterprises in its context (e.g., what is required from them) and to help the decision makers to design and refine their business strategies, and the responsibles for developing the EA to understand what is required from the resulting IS. Far from simple, the construction of Context Models (CM) is usually cumbersome, mainly due to the communication gap between the technical personnel (e.g., internal and external consultants) with a limited knowledge about the enterprise structure, operations and strategy, and its administrative counterpart, which impose pressure and time limitations to the process.

In order to deal with this problems, in the last years, the *i** notation has been intensively used to bridge the gap among technical consultants and non-technical stakeholders [2], and the DHARMA Method has been proposed [3], to discover enterprise architectures starting from the construction of CM expressed in *i** notation.

Furthermore, in [4, 5], a set of patterns and instantiation rules have been proposed, in order to semi-automatically construct CM. Although these proposals are useful in practice, they are based on syntactic techniques which constraint their practical application (e.g., because of the difficulty to identify terms such as synonyms and antonyms). To address this difficulty, this paper proposes to extend early phases of the DHARMA method with the incorporation of semantic technologies, aimed to improve the identification of elements and the construction of new CM, based in a semantic repository of CM elements. Since *i** CM will be defined through an ontology, this allows to keep a common vocabulary to share knowledge, and to use the capability of reasoning discovering relations existing among different models, making easier for Enterprise Architects to reuse experiences.

This paper describes the process used in the construction of the semantic repository of CM expressed in *i** notation. A short introduction to the *i** notation and the DHARMA method is presented in Sect. 2, the process of construction of CMs is exposed in Sect. 3 and then, the approach used in the process of creating the semantic repository is described in Sect. 4. Finally, a set of conclusions and future works are presented in Sect. 5.

2 Background and Related Problems

2.1 *The* i* *Framework*

The *i** framework [6] was formulated for representing, modeling and reasoning about socio-technical systems. Its modeling language is constituted by a set of graphic constructs which can be used in two models: the *Strategic Dependency* (SD) model, which allows for the representation of organizational actors, and the

Fig. 1 Excerpt of a CM representing the intentionality between actors *Organization* and *Customer*

Strategic Rationale (SR) model, which represents the internal actor's rationale. Since this work makes intensive use of SD models, we focus the explanation on its constructs.

Actors in SD models are represented by a circle. They can be related by is-a (subtyping) relationships and may have social dependencies. A *dependency* is a relationship among two actors, one of them, named *depender*, who depends for the accomplishment of some internal intention from a second actor, named *dependee*. The dependency is then characterized by an intentional element (*dependum*) which represents the dependency's element. The primary intentional elements are (see Fig. 1): *resource* represented by a rectangle (e.g., *Invoice* or *Voucher*), *task* represented by a hexagon (omitted in this work because considered to prescriptive for an early phase in the requirements engineering process), *goal* represented by an oval (e.g., *Purchases Invoiced*) and *soft-goal* represented by a shrunken oval (e.g., *Timely Payment*). Goals represent services or functional requirements, whilst soft-goals represent goals that can be partially satisfied or that requires additional agreement about how they are satisfied. Soft-goals are usually introduced to represent non-functional requirements and quality concerns. Resources on the other hand, represent physical or logical elements required to satisfy a goal whilst tasks specific ways to achieve them.

2.2 The DHARMA Method

The DHARMA (*Discovering Hybrid Architectures by Modelling Actors*) method, aims at the definition of enterprise architectures using the *i** framework. The strategic framework is based on Porter's model of *market forces* and *value chain* [7]. The first is designed to analyze the influence of five competitive forces on business context (*threat of new entrants*; *threat of substitution*; *bargain power of customers*; *bargain power of suppliers*; and *rivalry among current competitors*) and reason about strategies to make organizations profitable. To balance the forces, enterprises adopt an internal organization known as *Value Chain*, which encompasses five primary (*inbound logistics*; *operations*; *outbound logistics*; *marketing & sales*; and *support*) and four *support* (*Infrastructure*; *human resources* *management*; *technology development*; and *procurement*) value activities (VA), required to generate value and eventually a *margin* (difference among total value generated and cost of performing VA). Primary activities are core and specific of business whilst supporting ones are transversal to them. The DHARMA method has been used in several industrial case studies, which have let to the identification of improvement opportunities, some of which are addressed in this paper. The method is structured with four main activities:

Activity 1: Modelling the enterprise context. The organization and its strategy are carefully analysed, to identify its role inside the context. This analysis makes evident the *Context Actors* (CA) and the *Organizational Areas* (OA) which structure it (DHARMA recognizes four types of actors, human, organizational, hardware or software). CA are identified in relation to market forces and examined in relation to each OA in the value chain, to identify strategic needs among them (*Context Dependencies—CD-*). Also OAs are analysed in relation to each other to identify their strategic interactions (*Internal Dependencies—ID-*). *i** SD models are built and used to support reasoning and represent results from this activity. Various CM are constructed from the perspective of each OA, including their related CA and OAs as well as their CD and ID. Resulting models are eventually combined into a single Enterprise Context Model.

Activity 2: Modelling the environment of the system. An IS-to-be is placed into the organization (it can be a pure IS or a hybrid system including hardware, software or hardware with embedded software components) and the impact that it has over the elements in the CM is analysed. Strategic dependencies identified in the previous activity (internal and context), are examined to determine which of them may be totally or partially satisfied by IS. These dependencies are redirected inside the *i** SD diagram to the IS. The model includes the organization itself as an actor in IS environment, its needs are modelled as strategic dependencies over the IS.

Activity 3: Decomposition of system goals and identification of system actors. Dependencies included in the IS CM are analysed and decomposed into a hierarchy of goals required to satisfy them. The goals represent the services that IS must provide, to support interaction with CA and OA activities. An *i** SR diagram for the system is built, using means-end links of type goal-goal (representing then a decomposition of objectives into sub-objectives).

Activity 4: Identification of system architecture. Finally, goals included in the SR model are analysed and systematically grouped into *System Actors* (SA). Objectives are clustered into services, according to an analysis of the strategic dependencies with the environment and an exploration of software components marketplace. Relationships between SA that form IS architecture are described according to the direction of the means-end links that exist among the objectives included inside them. SA are not software components; they represent atomic software domains for which several situations may occur: there can be a software component covering the functionality of several SA; the functionality of a single SA is covered by several software components for ubiquity reasons e.g. mobile and local applications; or there can be cases for which no software components exist, leading to the need of bespoke software.

2.3 Scenario or Problem

Despite the guidelines proposed in [3], the construction of CMs is usually manual and ad hoc process. However, in [4, 5] some semantic alternatives are presented, which allows to reuse CM elements (actors and dependencies), contained in patterns. These patterns are used to support the construction of *i** SD CM from scratch and eventually to automate this process. Specifically in [5] some ortogonal dimensions have been identified for classifying generic actors (as example see Table 1, in relation to the generic actor *Customer*). Each dimension has a set of associated value labels, representing potential actor instances. Those labels have a set of generic dependencies associated to them. Based on this table, practitioners (systems engineers and administrative staff) can systematically identify a large number of actors in their operational context, selecting and combining labels from each dimension. To illustrate the approach, let's consider the first two labels of three of the *Customer's* categorization dimensions in table 1, *frequency/volume*, *distribution channel,* and *payment method.* In this case, 12 combinations representing potential instances of actors in the context of the organization are possible: *Potential Wholesaler Credit, Potential Wholesaler Cash, New Wholesaler Credit, New Wholesaler Cash, Important Wholesaler Credit, Important Wholesaler Cash, Potential Retailer Credit, Potential Retailer Cash, New Retailer Credit, New Retailer Cash, Important Retailer Credit, and Important Retailer Cash.*

Suppose that in a particular case, the *New Wholesaler Credit Customer* has been selected from this set of combinations, then all dependencies associated to its labels will be the potential dependencies to be included in the organizational CM. Thus, the identification of dependencies can also be automated. Although this approach constitutes a significant contribution, it has a problem with the labels and associated dependencies because those are "static", which introduces an important problem from the semantic point of view, for example by ignoring synonymous labels to the defined in the table. So, if the word "client" appears instead of "customer", the label wouldn't be recognized and therefore the potential dependencies won't be included in the CM.

2.4 Related Works and Literature Review

The use of ontologies has increased in a wide range of areas (including IS) in the last years and since this work focuses in CMs obtained as result of early requirements modeled according to the DHARMA Method activities, we have performed a mapping study of ontologies, which allow the semantic annotation of *i** models, obtaining few related works. In [8, 9] the development of OntoiStar and OntoiStar + is presented, both are meta-ontologies used to describe *i** models, aiming the integration of different *i** variants; based on these ontologies some related works has been performed, for example the development of a Tool for the

Table 1 Customer Dimensions and associated Dependencies

Generic actor	Dimension	Actor instances	Associated dependencies	Type	Direction
Customers	Frequency or volume	Potencial	Widespread promotions	Goal	→
			Promocional samples	Resource	←
		New	Membership card provided	Goal	→
			Special introduction prices provided	Soft goal	→
			Membership card	Resource	→
			Personal information registered	Goal	←
		Important	VIP benefits granted	Goal	→
			Personalized attention	Soft goal	→
			VIP card	Resource	→
			Important high volume order placed	Goal	←
	Distribution channel	Wholesaler	Product availability guaranteed	Goal	←
			Product distribution agreement signed	Soft goal	←
			Increase sales through the distribution chain	Soft goal	←
			Product distribution agreement	Resource	←
			Product distribution chain achieved	Soft goal	→
		Retailer	Restocking in small quantities provided	Goal	→
			Approach consumers through an specific location	Soft goal	←
			Increase sales through individual stores	Soft goal	←
		Specific market Segment	Specialized customer service infrastructure	Soft goal	→
			Trained stuff for specific needs	Soft goal	→
			Specific documents	Resource	→
	Payment method	Credit	Deferred payments	Goal	→
			Credit flexibility	Soft goal	→
			Acceptance of various credit cards	Soft goal	→
			Voucher	Resource	→
			Warranty documents	Resource	←
		Cash	Cash rebates	Goal	→
			Money	Resource	←

(continued)

Table 1 (continued)

Generic actor	Dimension	Actor instances	Associated dependencies	Type	Direction
			Technology, products or services provided	Goal	←
			Timely payments	Soft goal	←
			Products, services, technology	Resource	←
			Invoiced purchases	Goal	→
			Quality of products or servi ces	Soft goal	→
			Bill	Resource	→

Automatic Generation of Organizational Ontologies (TAGOON+) [9]; and a method for integrating the constructs of *i** variants through the use of ontologies [10].

In [11] the development of an Intermediate Retrieval Ontology is proposed, it allows the annotation of *i** models and their validation through the rules defined in the ontology, and eventually find domain ontologies useful to extend the vocabulary of the model under analysis. Even though that works allows the annotation of *i** models, this paper goes beyond, looking for the creation of a Semantic Repository of Context Models, from which a system could infer and thus help informatics consultants to perform enterprise analysis easier and faster, through the reuse and inference provided by the repository.

3 The DHARMA Method Application

The outcome of each DHARMA Method activity are a group of *i** models. Those models can be used as inputs to supply information to a semantic repository, in which the models could be represented through ontologies. This work focuses in the first two activities of the DHARMA Method.

The CM resulting from the first activity can be expressed in tabular form, using a structured table composed by the fields (see Table 2): *Actor1, Actor2, Dependency, Type, Direction;* where *Actor1* and *Actor2* can perform the role of *Depender* or *Dependee*,

Table 2 Tabular representation of the CM included in Fig. 1

Actor 1	Actor 2	Dependency	Type	Direction
Customer	Organization	Technology, products or services acquired	Goal	>
Customer	Organization	Technology, products or services	Resource	>
Customer	Organization	Quality of products and services	Soft goal	>
Customer	Organization	Invoice	Resource	>

based to the value of the *Direction* field. Whilst, the *Dependency* field describes the strategic interaction between *Actor1* and *Actor2*, and the *Type* field defines the type of the dependency: *Goal*, *Soft-goal*, *Resource* or *Task*.

The second activity of the method attempts to establish the dependencies which can be automated through an IS. From the tabular point of view, it's enough to extend the resulting table from the activity 1, including a pair of columns establishing if the system can cover *Partially* or *Totally* the services (functionality) required for satisfying the dependencies.

4 Semantic Repository Creation Process

In this section, we present a proposal for the creation of a semantic repository using the principles of Linked Data [12], departing from *i** models generated through the DHARMA method. The proposal is based on the guidelines defined by the W3C Working Group [13] and the guidelines proposed in [14].

4.1 Data Source Selection

In this section, the data source to be processed is identified. We used Excel files to represent the *i** models in tabular way, as mentioned in Sect. 3. These files have been collected over time in real experiences, in over 30 organizations where the DHARMA method has been used. In this paper, a first manual approximation is used to transform the raw data into RDF format.

4.2 URIs Definition

In this stage URIs representing resources are defined, allowing the unique identification of an *i** model. In W3C Working Group there are some guidelines for defining URIs for resources [15].

URIs design, the root URI is http://www.ucuenca.edu.ec/. The URI for the identification of resources (*i** model) modeled in the ontology is http://www.ucuenca.edu.ec/ontologies/iStar/<organization_name>#. The URI for the identification of each element consists of the project URI plus the description of the *i** element. For instance, in the case of actors, the *Customer* actor URI is defined as: http://www.ucuenca.edu.ec/ontologies/iStar/dgenerico#Customer; for dependencies and links the URI consists of the project URI plus the dependency description and the resource type, for instance http://www.ucuenca.edu.ec/ontologies/iStar/dgenerico#Timely_payments_Dependency and http://www.ucuenca.edu.ec/ontologies/iStar/dgenerico#Timely_payments_DependerLink for the link between

Table 3 Examples of the project URIs

Table elements	URI
Customer	http://www.ucuenca.edu.ec/ontologies/iStar/dgenerico#Customer
Organization	http://www.ucuenca.edu.ec/ontologies/iStar/dgenerico#Organization
Invoice	http://www.ucuenca.edu.ec/ontologies/iStar/dgenerico#Invoice_ Dependency
Resource	http://www.ucuenca.edu.ec/ontologies/iStar/dgenerico#Invoice_ Resource
Links between elements	http://www.ucuenca.edu.ec/ontologies/iStar/dgenerico#Invoice_ DependumLink
	http://www.ucuenca.edu.ec/ontologies/iStar/dgenerico#Invoice_ DependerLink
	http://www.ucuenca.edu.ec/ontologies/iStar/dgenerico#Invoice_ DependeeLink

the dependent and the dependency. Table 3 shows examples of URIs for each element in the last record of Table 2.

4.3 Vocabularies

According to [13], standardized vocabularies should be reused as much as possible to facilitate the inclusion and expansion of the web of data. Concerning to *i** vocabularies most authors describe their models and benefits. In the context of *i** model, we have selected the meta-ontology OntoiStar as the main schema for representing *i** models (SD and SR models). This meta-ontology allows modeling all *i** elements, both nodes (actor and dependencies) and relationships (dependencies, actors, etc.).

4.4 RDF Generation from i* Models

Some difficulties arise in the RDF generation of *i** models mainly due to the lack of similar experiences. Currently, there are several tools that can be used for RDF generation, however, an Excel table generated from *i** model has not been tested with these tools. Protege software was used as tool to generate RDF from *i** model. This process is manually performed in order to demonstrate the validity of our approach. The first step for generating RDF is to define a mapping between OntoiStar and CM elements of Tables 2 and 4 shows the mapping generated and Table 5 shows the ontology properties used as annotations.

For generating RDF, we used one generic *i** model and one related to a more specific enterprise business model; as a first approach, in order to test the generation

Table 4 Mapping between classes in OntoiStar and *i** elements

OntoiStar classes	*i** elements
Actor	Context actor (Depender, Dependee)
Dependency	Element representing a dependency
Goal/Soft goal/Resource/Task	Element representing the type of a dependency
DependerLink	Link between Dependers and Dependencies
DependeeLink	Link between Dependencies and Dependees
DependumLink	Link between Dependencies and its type

Table 5 Properties of OntoiStar used for the instantiation

Properties of OntoiStar
has_Dependency_DependerLink_source_ref
has_Dependency_DependerLink_target_ref
has_Dependency_DependeeLink_source_ref
has_Dependency_DependeeLink_target_ref
has_Dependency_DependumLink_source_ref
has_Dependency_DependumLink_target_ref
is_a

process, only the areas of the models in relation to the Customer actor and some of its dimensions (see Table 1) were considered. An example of the RDF resulting from the instantiation of an *i** dependency is shown following this paragraph. Due to space limitation, we will use Namespaces[1] to identify resources and properties.

Dependency: **Organization** depends on **Customer** to obtain **Timely Payments**
Namespaces:
PREFIX isd, **IRI** http://www.ucuenca.edu.ec/ontologies/iStar/dgenerico#
PREFIX ois, **IRI** http://www.cenidet.edu.mx/OntoiStar.owl#

```
isd:Timely_payments_DependerLink
ois:has_Dependency_DependerLink_source_ref isd:
Timely_payments_Dependency
isd:Timely_payments_DependerLink
ois:has_Dependency_DependerLink_target_ref isd:Organiza
tion
isd:Timely_payments_DependeeLink
ois:has_Dependency_DependeeLink_source_ref
isd:Timely_payments_Dependency
isd:Timely_payments_DependeeLink
ois:has_Dependency_DependeeLink_target_ref isd:Customer
```

[1]Namespace uniquely identifies a set of names, so that there is no ambiguity when objects having different origins but the same names are mixed together.

```
select ?DependumName ?DependeeName
where {
  ?s ois:has_Dependency_DependerLink_target_ref dg:Cliente .
  ?s ois:has_Dependency_DependerLink_source_ref ?Dependum .
  ?o ois:has_Dependency_DependeeLink_source_ref ?Dependum .
  ?o ois:has_Dependency_DependeeLink_target_ref ?Dependee .
  ?u ois:has_Dependency_DependumLink_source_ref ?Dependum .
  ?Dependum rdfs:label ?DependumName .
  ?Dependee rdfs:label ?DependeeName .
} ORDER BY ?DependumName
```

DependumName	DependeeName
"Invoice"@en	"Organization"@en
"Invoice"@en	"CAPEDI"@en
"Products"@en	"CAPEDI"@en
"Products acquired"@en	"CAPEDI"@en
"Quality of products and services"@en	"Organization"@en
"Quality services"@en	"CAPEDI"@en
"Technology products or services"@en	"Organization"@en
"Technology products or services acquired"@en	"Organization"@en

Fig. 2 SPARQL query of actors and dependencies related to customer actor

```
isd:Timely_payments_DependumLink
ois:has_Dependency_DependumLink_source_ref
isd:Timely_payments_Dependency
isd:Timely_payments_DependumLink
ois:has_Dependency_DependumLink_target_ref
isd:Timely_payments_Softgoal
```

The annotation process is performed for all dependencies included in the CM. As a result we obtain a semantic repository of *i** models in OWL format, which are stored in a Triple Store[2] for future exploitation.

4.5 *Data Linking*

As shown in Sect. 2.4 (and as far as we know), there are not similar Linked Data approaches; therefore, it is not possible to link models defined in this work with models in other repositories.

4.6 *Publication and Exploitation*

The semantic data generated (RDF) from *i** models was loaded into the Virtuoso[3] triple store, which allows the storage of RDF. Based on this knowledge, SPARQL[4] queries (SQL-style queries) can be performed. An example of SPARQL query to obtain *Dependencies* and *Dependees* from a given *Depender*, in this case the Depender is the Customer actor is shown in Fig. 2.

---

[2]Triple Store: Purpose-built database for the storage and retrieval of triples through semantic queries.

[3]Virtuoso: https://www.w3.org/2001/sw/wiki/OpenLink_Virtuoso.

[4]SPARQL: Query language for RDF. http://www.w3.org/TR/rdf-sparql-query/.

5 Conclusions and Future Work

This paper presents a first idea for generating a semantic repository of *i**-based CMs, following the best practices of Linked Data. Based on the selected ontology model, we proceeded to instantiate the ontology using data obtained from 30 industrial experiences. The data was uploaded into a semantic repository which allows its analysis through SPARQL queries, and eventually infer new *i** CM, based on the properties of the ontology.

As future work we plan to extend this work, starting for the implementation of an automatic process to transform the data contained into Excel sheets to RDF triplets, discover links between the dataset and DBPedia through Silk framework, and sharing the information included in the 30 CM stored both in tables and the Triple Store, in order to provide Open Linked Data, which could be accessed for everyone. In addition, we pretend to extend this work to consider the two additional activities of the DHARMA Method, attempting to automate the inference of architectonic models of IS. In the same way, we pretend to exploit the semantic data generated through model SPARQL queries, which allows the extraction of similar patterns between different *i** models instantiated in the repository. Also, we planned to link the OntoiStar ontology with Domain ontologies, this will allow the semantic enrichment of *i** models and to create recommendation systems based on the knowledge registered in the repository.

References

1. The Open Group: The Open Group Architecture Framework (TOGAF) version 9 (2009)
2. Carvallo, J.P.: Supporting organizational induction and goals alignment for COTS components selection by means of *i**. ICCBSS (2006)
3. Carvallo, J.P., Franch, X.: On the use of *i** for architecting hybrid systems: a method and an evaluation report. PoEM (2009)
4. Carvallo, J. P., Franch, X.: Building strategic enterprise context models with *i**: a pattern-based approach. In: 5th International *i** Workshop (iStar'11) (2011)
5. Abad, K., Carvallo, J.P., Peña C.: iStar in practice: on the identification of reusable SD context models. In: 8th International *i** Workshop (iStar'15) (2015)
6. Yu, E.: Modelling strategic relationships for process reengineering (1995)
7. Porter, M.: Competitive Strategy. Free Press, New York (1980)
8. Najera, K., Martinez, A., Perini, A., Estrada, H.: Supporting *i** model integration through an ontology-based approach. In: 5th International *i** Workshop (iStar'11), pp. 43–48 (2011)
9. Najera, K., Martinez, A., Perini, A., Estrada, H.: An ontology-based methodology for integrating *i** variants. In: Proceedings of the 6th International *i*Star Workshop (*i*Star 2013) (2013)
10. Vazquez, B., Estrada, H., Martinez, A., Morandini, M., Perini, A.: Extension and integration of *i** models with ontologies. In: Proceedings of the 6th International *i*Star Workshop (iStar 2013) (2013)
11. Beydoun, G., Low, G., Gracía-Sanchez, F., Valencia-García, R., Martínez-Béjar, R.: Identification of ontologies to support information systems development
12. Berners-Lee, T.: Linked Data. http://www.w3.org/DesignIssues/LinkedData.html (2006)

13. W3C Working Group: Best Practices for Publishing Linked Data. http://www.w3.org/TR/ld-bp/. (2014)
14. Villazon, B.: Best Practices for Publishing Linked Data (2011)
15. W3C Working Group: Best Practices URI Construction. http://www.w3.org/2011/gld/wiki/223_Best_Practices_URI_Construction (2012)

Operations Research Ontology for the Integration of Analytic Methods and Transactional Data

Edrisi Muñoz, Elisabet Capón-García, Jose M. Laínez-Aguirre, Antonio Espuña and Luis Puigjaner

Abstract The solution of process systems engineering problems involves their formal representation and application of algorithms and strategies related to several scientific disciplines, such as computer science or operations research. In this work, the domain of operations research is modelled within a semantic representation in order to systematize the application of the available methods and tools to the decision-making processes within organizations. As a result, operations research ontology is created. Such ontology is embedded in a wider framework that contains two additional ontologies, namely, the enterprise ontology project and a mathematical representation, and additionally it communicates with optimization algorithms. The new ontology provides a means for automating the creation of mathematical models based on operations research principles.

Keywords Operations research · Enterprise wide optimization · Decision support systems · Knowledge management

E. Muñoz (✉)
Centro de Investigación en Matemáticas A.C., Jalisco S/N, Mineral y Valenciana, 36240 Guanajuato, Mexico
e-mail: emunoz@cimat.mx

E. Capón-García
Department of Chemistry and Applied Biosciences, ETH Zürich, 8093 Zurich, Switzerland
e-mail: elisabet.capon@chem.ethz.ch

J.M. Laínez-Aguirre
Department of Industrial and Systems Engineering, University at Buffalo, 14260 Amherst, NY, USA
e-mail: jmlainez@gmail.com

A. Espuña · L. Puigjaner
Department of Chemical Engineering, Universitat Politècnica de Catalunya, Av. Diagonal, 647, E08028 Barcelona, Spain
e-mail: antonio.espuna@upc.edu

L. Puigjaner
e-mail: luis.puigjaner@upc.edu

J. Mejia et al. (eds.), *Trends and Applications in Software Engineering*, Advances in Intelligent Systems and Computing 405,
DOI 10.1007/978-3-319-26285-7_12

1 Introduction

Process industries are highly complex systems consisting of multiple business and process units, which interact with each other, ranging from molecule to enterprise level. Therefore, in order to solve real world problems, it is necessary to develop algorithms and computational architectures so that large-scale optimization models can be posed and solved effectively and reliably. Hence, the collaboration among different scientific disciplines, namely process systems engineering, operations research and computer science, is highly important [1]. In this sense, enterprise-wide optimization (EWO) is a discipline related to the optimization of supply operations, manufacturing and distribution in a company [1]. A challenge in EWO consists in developing flexible modeling environments for the problem representation, which is the ultimate basis for reaching efficient decision-making. A key feature in EWO is the integration of information and decision-making along the various functions and hierarchical levels of the company's supply chain. Current information technology tools, such as data mining, allow a high degree of information flow in the form of transactional systems (e.g. ERP's) and efforts have been recently devoted to integrate transactional systems and analytical models (e.g. optimization and simulation). These transactional systems must interact with analytical models providing the necessary data to reach appropriate solutions. However, further development is still necessary to easily develop, build and integrate different enterprise models. Such integration should reflect the complex trade-offs and interactions across the components of the enterprise.

This work proposes the creation of an Operations Research (OR) semantic model as a step forward in capturing the nature of problems and technologies for decision making in the enterprise. Specifically, the whole process of decision making and the creation and classification of equations according to their structure are qualitatively represented in terms of OR principles. This OR ontology (ORO) is integrated with other two semantic models previously developed, namely the Enterprise Ontology Project (EOP) [2] and the Ontological Math Representation (OMR) [3], thus enhancing the functionalities of the original ontological framework (Fig. 1). The scope of these models comprises the representation of the real system for EOP, the mathematical representation domain for OMR, and finally the problem design representation for ORO.

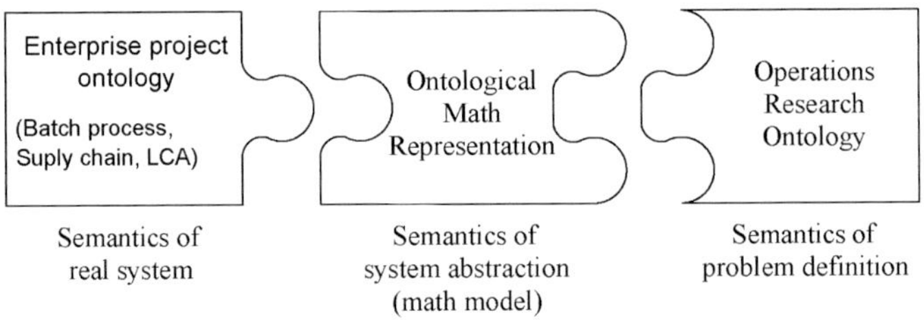

Fig. 1 Relationship among the semantic models with the represented domain

1.1 Operation Research

The Operations Research as science, provides analytical models (technical and mathematical algorithms) to solve decision problems. Operations Research as an art, provides creativity and ability of problem analysis, data collection, model building, validation and implementation for the scientific achievement of problems solution. Besides, the establishment of communication lines between sources of information and analytical models, is a key factor of success.

Thus, the decision making process involves building a decision model and finding the optimal solution. In this way, the problem model is defined as an objective and a set of constraints that are expressed in terms of decision variables of the problem. Decision models allow the identification and systematic evaluation of "all" the possible choices for problem solution. The basic components of an operations research system are decision choices, problem constraints and the objective criteria. The resulting system is an abstraction of the reality which considers the identification of key factors, such as, variables, constraints and parameters. We can reach feasible, unfeasible, optimal and suboptimal solutions, when a problem is solved. The quality of the optimal solution depends on the set of feasible options that are found by the problem modeling and problem solution process. There are different solution techniques such as linear programming (LP), integer programming (IP), dynamic programming (DP), mix integer linear programming (MILP), non linear programming (NLP), and mix integer non linear programming (MINLP) representing algorithms to solve particular classes of operations research models.

Models need to be supported by reliable data in order to provide useful information to the decision maker. Data collection may be the most challenging part of the model to be determined. While modeling experience accumulates, a key factor is the development of means to collect and document data in a useful way. If the data accommodates probabilistic data, then probabilistic or stochastic models are derived.

Basically the process for finding a solution (calculation method) in operations research for mathematical models is iterative in nature. Indeed, the optimal solution is reached by steps/iterations in which each iteration results in feasible and unfeasible solutions.

The phases of a study of operations research are: (i) problem definition; (ii) model building; (iii) model solution; (iv) validation of the model; and, (v) implementation of optimal results. *The problem definition* comprises three main aspects: (a) goal or objective definition; (b) decision alternatives identification; and, (c) identification of the limitations, restrictions and system requirements. *Model building* specifies quantitative expressions which defines the objective and constraints based on system variables. *The model solution* depends on the mathematical technique to be used. Additionally, it is recommended to implement a sensitivity analysis to study the behavior due to changes in the parameters of the system. *The validation of the model* consists of the adequate description of the real system by means of the created model by using historical data. Finally, *the implementation of optimal results* involves translating the results into detailed operating instructions.

2 Operations Research Ontology

Moreover, ontologies are emerging as a key solution to knowledge sharing in a cooperative business environment [4]. Since they can express knowledge (and the relationships in the knowledge) with clear semantics, they are expected to play an important role in forthcoming information-management solutions to improve the information search process [5]. Thus, a development of operations research semantic model which aims to support a formal and systematic process for problems model and design. In order to design a robust and reusable model, a systematic methodology based on a continuous improvement cycle has been followed [6]. This methodology is composed by 5 main steps, which are, plan, do, check, act and re-plan phases. Next, the main characteristics of the above mentioned phases are briefly described.

2.1 Plan Phase

Operation research models support a rigorous problem description using mathematical language (symbols, variables, equations, inequalities, etc.), by structuring the problem as a set of equations. The set of equations represent the objective of the problem and different restrictions and constraints of the system where the problem is found. The proposed semantic model aims to formally represent operations research models for engineering systems, giving special importance to the model construction, structuring and solution steps. Several sources of knowledge have been consulted [7, 8], and the result of the formalization has been validated with engineers currently involved in operations research and its application to the area of process systems engineering.

The resulting semantic model must formalize the design of any operations research problem (Fig. 2), allowing to:

- support the correct structure of operations research, objective function and constraints.
- formalize the meaning of system equations, simple equations and elements within those equations.
- formalize the relationship between system of equations, equations and elements of mathematical models.
- share knowledge with other ontologies for reaching a better system understanding.

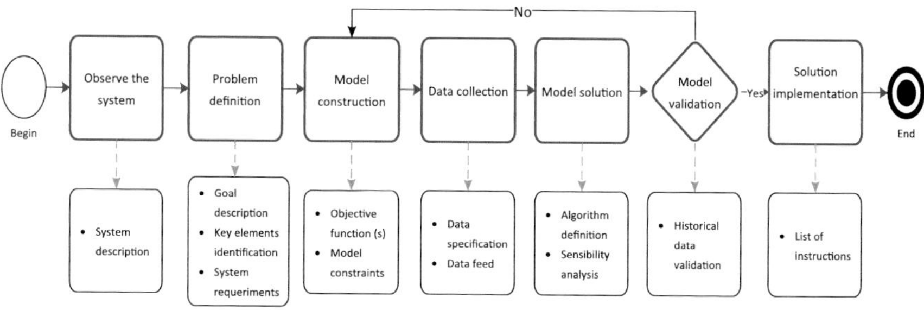

Fig. 2 Flow diagram of the operations research process

- analyze more efficiently the effects of changes in the problem when factors affect the system.
- strengthen decision-making based on operations research modeling.

2.2 *Do Phase*

The creation of a general operations research modeling ontology was performed using the principal components of the operations research domain gathered from the sources of knowledge and listed in a glossary of terms. Concepts such as Activity, Algorithm, Allocation, Alternatives identification, Assumed real system, Attribute, Break-even analysis, Collection of information, Cost-benefit analysis, Decision trees, Decision-making, Efficiency, Feasible solution, Goal setting, Implementation, Linear programming, Mathematical model, Model, are defined in Table 1, which presents the complete list of terms considered in this work related to the operations research domain.

On the basis of the above presented list, the concepts required to construct the operational research model results in ontological classes which are structured in a taxonomy, as shown in Fig. 3, representing the top classes of the model. These classes are the main building blocks for representing operations research as base for a robust decision making process. On the other hand, some verbs used in the operations research, such as, performs, achieves, has solution relates to, satisfies, establishes, specifies, among others, were used in order to stablish object properties.

The ontology was implemented in Protégé [9] using the Web Ontology Language (OWL). A key point of this model is its connectivity to the system reality. Such connection is achieved by means of the class “Term”, the modeling entity directly related to the engineering domain to be mathematically modeled. Quantitatively speaking, the ontological model contains 72 classes, 52 properties and 9 axioms (see Fig. 3).

Table 1 Basic concepts regarding operations research ontology

Concept	Description
Algorithm	Method to solve a problem through a defined, precise and finite set of instructions
Allocation	The activity of allocate the necessary mount of resources to each activity needed to develop it
Assumed real system	It is an abstraction of the situation resulting from the identification of key factors (variables, constraints and parameters)
Break-even analysis	Analysis of relationships to determine the size or transaction volume reaches a balance between losses and profits
Cost-benefit analysis	A method for finding the best result between benefits and costs, expressed in monetary value
Decision trees	A technique to analyze sequential decisions based on the use of outcomes and associated probabilities
Decision-making	The activity of select certain actions from several options rationally
Efficiency	Achievement of objectives with the least amount of resources
	Achievement of objectives at the lowest cost or other unintended consequences
Feasible solution	A set of values of the problem variables that satisfies the equations
Implementation	It implies the translation of the results into detailed operation instructions
Linear programming	It is mathematical algorithm or procedure by means of which a problem, formulated in terms of lineal equations, is solved optimizing an objective function, also lineal
Mathematical model	Is a model where the objective function and the constraints are expressed in quantitatively or mathematical way
Model	It is an abstraction of the assumed real system that identifies relevant relationships of the system in the form of an objective function and a set of constraints
Model construction	It specifies quantitative expressions for the objective function and the constraints in function of the constraints variables
Model validation	It is presented when the performance of the model is similar to the performance of the real system by using historical data
Operations research	A scientific discipline, which applies theory, methods, and special techniques in order to search the solution to decision making problems related to the conduction and coordination of operations (or activities) within an organization
Optimal solution	A feasible solution that optimizes the objective function
Optimization	Optimization attempts to answer a problem choosing the best among a set of items
Phase	The problem creation is divided in compulsory phases, which are mainly situation examination and information collection
Problem definition	It is defined by three main tasks: (1) Goal description; (2) Identification of the system decision choices; and (3) Identification of limitations, constraints and requirement of the system
Model solution	It depends of the complexity and nature of the model. For the solution some mathematic, simulation or heuristic, or any combination must be

(continued)

Table 1 (continued)

Concept	Description
	used. Even more, some sensitive analysis should be performance in order of study the behavior of the parameters in the system
Risk analysis	Analysis that weighs the risks of a situation to include chances to get a more accurate assessment of the risks involved.
Sensitive analysis	Methodology that allows you to examine the behavior of a result concerning controlled variations of a set of independent variables
	Calculation of the effect of an exogenous variable produces in another variable
Simulation model	It is a model where the system is expressed in basic or elemental modules and integrated bye the "if/then" logic relations
Stochastic model	It is a model where the data is unknown and probabilistic approaches are used in order to obtain data
Suboptimal solution	A feasible solution which is not the optimal solution
System	It is a set of parts or related items organized and interacting to achieve a goal. The systems receive (input) data, energy or environmental matters and provide (output) information, energy or matter
Unfeasible solution	A set of values of the problem variables that do not satisfy the equations

2.3 *Check Phase*

The language and the conceptuality were checked by experts in the domain to ensure that the model matched the users' requirements for the design and development of operations research models. Basically, the ontology was checked by members of the research groups specialized in optimization approaches applied to process system engineering (two professors, three associated researchers, two Ph.D. students and two informatics engineers). In addition, the consistency of the model was checked using ontology reasoners, such as, Pellet Incremental and Hermite reasoners from Protégé. The reasoning time for checking the consistency of the classes and of the model was 0.624 and 0.764 CPU's, respectively, in an intel-core 2 at 2.83 GHz, resulting in a successful compilation. Finally, the checking and validation of the ontological framework consisted in verifying that the operations research models could be successfully instantiated in the ontology. This step comprised a preliminary approach to check the usability of the model.

2.4 *Act Phase*

This phase comprised all the actions on the model necessary to repair defects and implement suggestions made during the previous phase. Specifically, as an

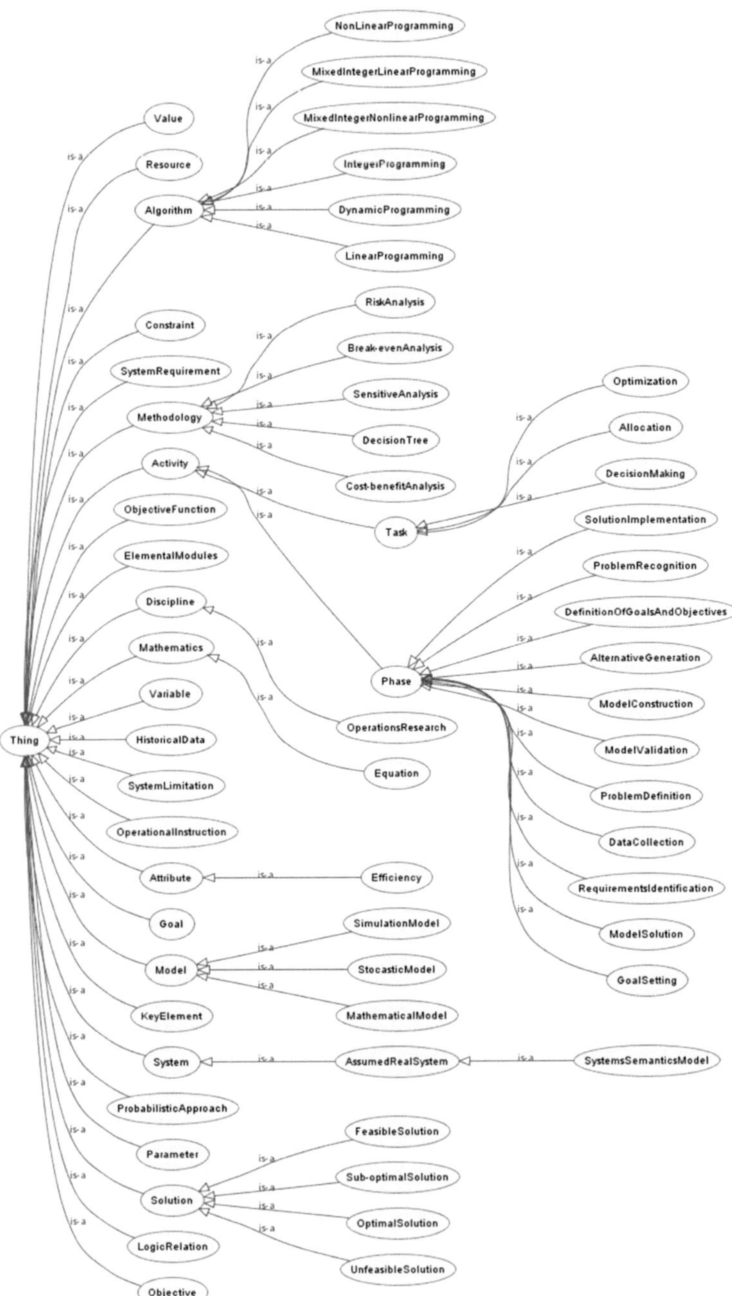

Fig. 3 Extract of the taxonomy of the operations research ontology

example, the use of the class "Element" was appropriately tuned, in order to specify that the real meaning regards a "Mathematical Element" and unnecessary classes and properties were eliminated after debugging. Finally, the use of the current model and all the formal changes made to the ontological model were recorded in the project documentation.

2.5 *Re-planning Phase*

The results of this phase are included in the previous subsections, which present the final structure of the pursuit project. However the project has been re-planned several times adding new functionalities to the operations research domain. The whole final semantic model can be inspected and downloaded at http://bari.upc.es/cgi-bin/samba/smbdir.pl?group=&master=&host=neo&share=guest03&dir=\ORO&auth=OT8COzZ,fGcDEionFyQjFg8=.

3 Proposed Framework

In this section, the procedure of the whole framework is presented. As a first step, the actual state of the process is captured by the instantiation of the Enterprise Ontology Project which contains an integrated representation of the enterprise structure, ranging from the supply chain planning to the scheduling function, thus comprising activities related to the operational, tactical and strategic functions. As a next step, by means of the instantiation of the Ontological Math Representation, the various mathematical models (mathematical elements) already established or the design of new ones are translated into a semantic representation as mathematical expressions in order to capture the mathematical meaning of enterprise domain elements. This model relates existing classes belonging to the enterprise domain ontology to mathematical elements to understand and translate the system abstraction in equations. Finally, this work proposes the Operations Research Ontology in order to, on the one hand, formalize and support the processes of: (i) problem abstraction, (ii) analytical model building, (iii) problem solution, (iv) verification and (v) deployment of the best solution; and on the other hand, allow automated representation of the results of this whole process in standardized formats (e.g., the so-called mathematical programming formats MPS and NL).

The workflow diagram of the proposed framework is illustrated in Fig. 4. The first phase aims at reaching a formal conceptualization of the real system of the process industry under study. This step encompasses the standardized semantic description of the system using the enterprise ontology project (EOP), and the definition and acquisition of the required dynamic and static data. The second phase pursues the future formalization of the mathematical equations describing the system abstraction using the ontological mathematical representation (OMR), this

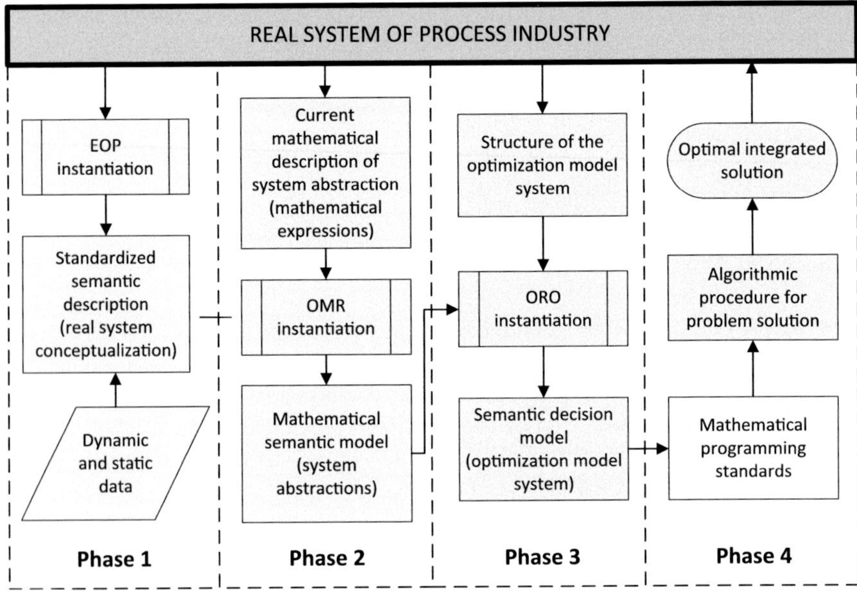

Fig. 4 Flow diagram of the design of problem system construction framework

phase results in a potential mathematical description of the entire system. The OMR can capture both mathematical expressions already in use and new developments for the system conceptualization. Thirdly, the structure of optimization model system along with the mathematical semantic model are the basis for instantiating the operations research ontology (ORO) and obtain a semantic decision model for optimization purposes. ORO has the task of designing the structure and the equation system in order to define a certain problem following the operations research guidelines. Finally, the mathematical programming standards are applied to the semantic decision model and the problem is solved to reach the optimal integrated solution, which assists managers in making the decisions to be deployed in the real system.

4 Case Study

To demonstrate the functionalities of this integrated ontological framework, a case study presented by Muñoz et al. [2] is considered. It consists of an integrated supply chain network planning and scheduling problem. In this work, the processes encompassing the problem definition, the problem formulations, as well as the solution procedure are represented within the "Operations Research Ontology". Therefore, the whole process of operations research has been applied to the problem solving supported by the knowledge management framework.

Table 2 Extract of "Problem definition" in the case study

Goal definition	Key elements identification
Maximize the economic performance of the whole supply chain structure	Direct cost parameters, such as production, handling, transportation, storage, and raw materials
Determine the assignment of manufacturing and distribution tasks to the network node	Indirect expenses, associate with the capacity utilization
Determine the amount of final products to be sold	Prices of the final products in the markets
Determine the amount of transported material among facilities	Set of suppliers with limited capacity provide raw materials to the different production plants
Determine the detailed batching, sequencing and timing of tasks in each production plant	Set of production plants and the distribution centers are located in specific geographical sites and provide final products to the markets
System limitations	System requirements
Mass balances have to be respected	Each production plant produces certain amounts of final products using equipment technologies which have defined installed capacity and minimum utilization rate
The final products are stored in one of the distribution centers before being sent to the markets	Maximum capacity limitations have to be considered for each treatment technology
Each market has a nominal demand of final products along a fixed time horizon	Available transportation links
A certain supply chain network structure	Production routes are defined in the product recipes, which contain mass balance coefficients and the consumption of production resources
	Unfulfilled demand cannot derive in back orders

As a result of the "Observe the system" phase, the case study has been captured using of the Enterprise Ontology Project, which allows to instantiate all the features of the system, from the production process level to the whole supply chain network. The following phase concerning "Problem definition" aims to capture four key issues: (i) goal description, (ii) key elements identification, (iii) system limitations, and (iv) system requirements (Table 2).

The "Model construction" has been derived from the "Problem definition", which provides the elements for the model creation. Specifically, links among the four different issues identified in the "Problem definition" are established by means of a relation matrix, and the sets, parameters, variables and groups of equations are accordingly derived. For example, the goal "Determine the amount of final products to be sold" is related to the continuous variable SLsff't, which represents the sales of product s at time period t produced in facility f to market f'. Thus, the semantic modeling of the model is supported by the "Ontological Mathematical Representation", which is also related to the "Enterprise Ontology Project".

The “Data collection” stage concerns the relation of the different sets and parameters of the specific problem instantiation to their current value, which are usually stored in organized databases. Thus, the “Model solution” considers the algorithm selection according to the specific features of the mathematical model, and proceeds to the problem solution. Finally, the steps “Model validation” and “Implementation” consider the validation of the model according to historical data and decision-maker expertise, and the acceptance and application of the resulting solution for the real system.

5 Conclusions

The ontological framework provides a tool to build computational optimization models for the enterprise decision-making and to allow the comprehensive application of enterprise wide optimization throughout the process industry. This framework encompasses the steps of OR, and communicates with two previously existing semantic models related to enterprise wide and mathematical representation. This extended framework also allows the building of more accurate models for the chemical process industry and the full integration and solution of large-scale optimization models. The main contribution of this work is the systematic and rigorous representation of the whole decision-making process from the problem conception to implementation. As a result, creation and re-use of analytical models in the industry can be semantically supported, providing a higher flexibility and integration for model building of enterprise operations. Finally, Operations Research Ontology aims to support problem design and mathematical formulations (modeling & programming) for decision making in order to reach EWO by standardizing and facilitating the decision making process based on operations research, creating and classifying systems of mathematical equations according to the system’s specification and decision type and reusing knowledge from other ontology models (Enterprise Project and Ontological Math Representation).

Acknowledgments Authors would like to acknowledge the Spanish Ministerio de Economía y Competitividad and the European Regional Development Fund for supporting the present research by projects EHMAN (DPI2009-09386) and SIGERA (DPI2012-37154-C02-01). Finally the financial support received from CIMAT México is also fully acknowledged.

References

1. Grossmann, I.E.: Enterprise-wide optimization: A new frontier in process systems engineering. AIChE J. **51**, 1846–1857 (2005)
2. Muñoz, E., Capón García, E., Laínez, J.M., Espuña, A., Puigjaner, L.: Integration of enterprise levels based on an ontological framework. Chem. Eng. Res. Des. **91**, 1542–1556 (2013)

3. Muñoz, E., Capón García, E., Laínez, J.M., Espuña, A., Puigjaner, L.: Mathematical knowledge management for enterprise decision making. In: Kraslawski, A., Turunen, I. (eds.) Computer Aided Design: Proceedings of the 23rd European Symposium on Computer Aided Chemical Engineering, Lappeeranta Finland, pp. 637–642 (2012)
4. Missikoff, M., Taglino, F.: Business and enterprise ontology management with symontox. In: Heidelberg, S.B. (ed.) The Semantic Web—ISWC 2002, 2342/2002, pp. 442–447. (2002)
5. Gruber, T.R.: A translation approach to portable ontology specifications. Knowl. Acquis. **5**(2), 199–220 (1993)
6. Muñoz, E.: Knowledge management tool for integrated decision-making in industries PhD Thesis, Universitat Politècnica de Catalunya Poner Barcelona, Spain (2011)
7. Ravindran, A.R.: Operations Research and Management Science Handbook. CRC Press Taylor and Francis Group, New York (2007)
8. Taha, H.A.: Operations Research. Prentice Hall, New Delhi (1995)
9. Horridge, M., Jupp, S., Moulton, G., Rector, A., Stevens, R., Wroe, C.: A practicalguide to building owl ontologies using protege 4 and CO-ODE tools. Technical report. The University of Manchester (2007)

Further Reading

10. Smith, T.F., Waterman, M.S.: Identification of common molecular subsequences. J. Mol. Biol. **147**, 195–197 (1981)

How to Think Like a Data Scientist: Application of a Variable Order Markov Model to Indicators Management

Gustavo Illescas, Mariano Martínez, Arturo Mora-Soto and Jose Roberto Cantú-González

Abstract The growing demand for specialists in analyzing large volumes of data has led to an emerging profile of knowledge managers known as data scientists. How to address the different and complex scenarios with mathematical methods makes a difference when to apply them successfully in a dynamic environment such as the management indicators. For this reason, the authors present in this article a case study of prognostic indicators, developed in the field of finance, making use of mathematical Markov model which has prototyped in an abstract technological implementation with the capabilities to implement cases in other contexts. The purpose of the case study is to verify if the different levels of analysis of the Markov model provide knowledge to the prognosis by indicators while the application of the proposed methodology is shown. Thus, this work introduces to the threshold of a methodology that leads to one of the ways on how to think like a data scientist.

Keywords Data scientist · Markov chains · Indicators · Knowledge management · Forecast

G. Illescas (✉) · M. Martínez
Faculty of Exact Sciences, Computer Science Department, Universidad Nacional del Centro de la Provincia de Buenos Aires, Tandil, Argentina
e-mail: illescas@exa.unicen.edu.ar

M. Martínez
e-mail: mmartinez@slab.exa.unicen.edu.ar

A. Mora-Soto
Mathematics Research Center, Zacatecas, Mexico
e-mail: jose.mora@cimat.mx

J.R. Cantú-González
Department of Industrial & Systems Engineering, Systems School, Universidad Autónoma de Coahuila, Acuña, COAH, Mexico
e-mail: roberto.cantu@uadec.edu.mx

J. Mejia et al. (eds.), *Trends and Applications in Software Engineering*, Advances in Intelligent Systems and Computing 405,
DOI 10.1007/978-3-319-26285-7_13

1 Introduction

In a previous research work [1] authors presented a validation of indicators' forecasting methods for organizational knowledge management; as it is detailed before, authors decided to use extended Markov models to widen the scope of the analysis done in the previous work and to allow exploring the evolution of indicators that have previously registered on a database. Among the exploration parameters it is possible to define the extension order during the modeling of this method, this is known as variable order Markov model [2]. The methodologic proposal presented in this paper aims to be a novel alternative for Data Scientists (DS) by using an additional variable in the analysis space of an indicator's evolution in order to improve the estimation accuracy.

According to [3], a DS is a person who combines the knowledge of statistics, information technology and information design with strong communication and collaboration skills. The workplace of the person must allow self-provision of information, to enable it to be inquisitive and creative.

A proper methodology to analyze the information stored in a database includes the definition of indicators, same that have been used as powerful tools in organizational management and have leaded to the definition of the balance score card (BSC) [4]. Indicators emerged from the integration of different perspectives that are considered relevant for analysis, it includes the importance of selecting a time dimension that is fundamental for an evolutionary analysis of indicators; values for indicators are associated to a finite set of states (normally associated to the concept of semaphore) that lead to observe the evolution of the indicators' states. This evolution could be studied and analyzed using Markov models, in this sense, the contribution of this work is the possibility to vary the analysis windows in Markov models enriching the estimation possibilities and adding value for the data science realm.

In the following sections a brief introduction to the proposed methodology is given as well as a description of the case study including the selected indicators to validate authors' proposal. During the case study presentation an example of an implementation prototype is shown followed by the results and conclusions.

2 Development of Methodological Aspects

As depicted in Fig. 1 the proposed methodology includes of set of phases that have to be completed by a DS in order to use it. After the diagram, an explanation of each phase is presented. During the case study application, each application phases is shown, seeing how the management by indicators is done.

A. *Selecting a Mathematical Model.* This phase implies to define a case study related to event that want to be analyzed and it is strongly connected its context;

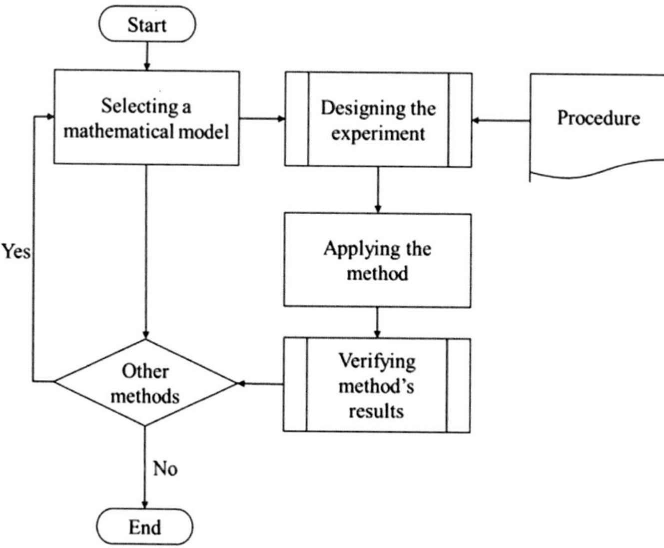

Fig. 1 Phases' flow for data scientists of the proposed methodology

this model selection is critical and it relies on the experience and expertise of the DS.

B. *Designing the Experiment in the Application Prototype.* Experiment design leads to the DS to successfully accomplish all the tests designed for the case study. In order to become more efficient during this phase an application prototype was implemented by authors to facilitate the selection of the analysis variable (the indicator) from an indicators' database. The DS has to select an analysis variable and to observe its behavior along the time series defined in the database by using the elements of traditional statistics such as average, mode, quartiles and variable dispersion [5]. Later the experimentation scenarios have to be defined by the DS base on the distribution of the observed values; at least two scenarios have to be defined, one to test or train the model and other to validate the results.
C. *Applying the Method.* This phase implies to apply the method in the case study as it is described in Sect. 3.
D. *Verifying Method's Results.* In order to complete this phase it is necessary to split the data according to the range of available dates and the observation period. For example, if the observation period of the variable is defined in months and you have historical data from five years (60 observations), it is possible to use 90 % of data for the first set (training or testing) and the rest of data for validation (6 observations).
E. *Evaluating with Other Methods or with a Modification of the Current Model.* In the case study depicted later in Sect. 3 the results from using Markov models are presented.

The aspects presented introduce us to the threshold of a methodology that leads us to one of the ways on how to think like a data scientist, sorting out the process of thinking and analysis in this case as a management tool by indicators.

3 Application on a Case Study

In this section authors present a case study where the methodological aspects defined in Sect. 2 we applied. The purpose of the case study is to verify if the different levels of analysis of the Markov model, provide knowledge to the prognosis by indicators while the methodology proposed application is being shown.

3.1 Phase A: Selecting a Mathematical Model

This section presents the theoretical framework used by a DS to develop a case study using Markov models. In this regard there are experiences on specific cases, like the use of Markov indicators and chains for decision making on new products [6]; simulation of indicators supported by Montecarlo Markov chains [7]; validation of the relation cause-effect through Bayes networks [8]. Despite the contributions that these jobs done, they do not end the difficulties which enclose in itself the management by indicators, including: the lack of validation mechanisms and the lack of integration of proven methods as part of methodologies (for instance the BSC) most of all, they show isolated successful cases [1]

A stochastic process is a set of random variables {Xt, t $\in$ T}. Normally t indicates time and all the possible values of X are identified as the process states along the time t. If T is discrete the process is time-discrete. Each one of the random variables of the process has its own probabilistic distribution function and could be correlated or not. Each variable or set of variables define a stochastic process, that usually describes a process that evolves in time [9]. Among the stochastic processes we can find the Markov Process, a special kind of discrete process where time's evolution only depends on the current state and not from the former ones [10]. Markov models imply random phenomena that are dependent of time where a specific property is true: a Markov chain that is in a state j at the time j + 1 depends only on the state it was in the instant n. The term Markov Chain is frequently used to define Markov processes on spaces of discrete estates (finite or numerable). In formal terms a Markov chain is a sequence of random variables {Xn, nN} that meets the following:

1. Each variable X_n takes values from a finite or numerable set E that is known as *states space*.
2. The sequence of variables verifies the condition or property of Markov.

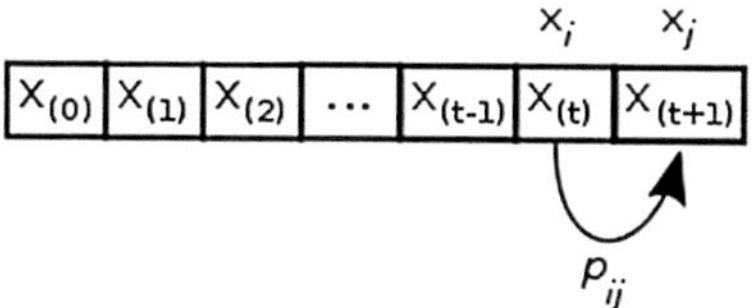

Fig. 2 Discrete-time Markov chain

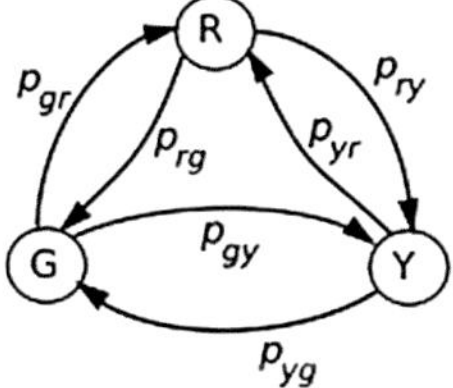

Fig. 3 State transition graph for a Markov chain

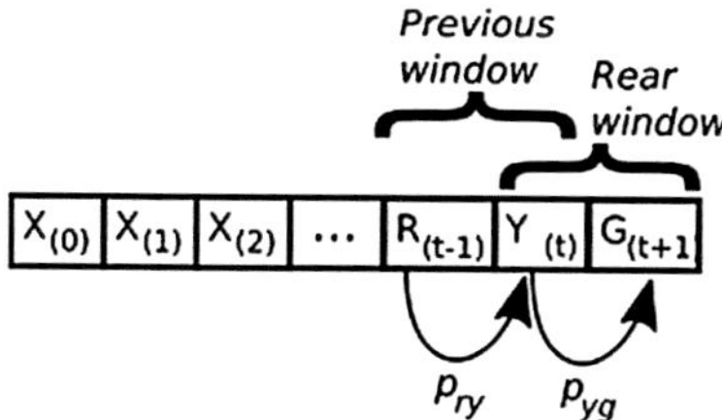

Fig. 4 Last three values of the chain as well as the analysis windows of two states

$$P(X_{n+1} = i_{n+1} | X_0 = i_0, \ldots, X_n = i_n) = P(X_{n+1} = i_{n+1} | X_n = i_n) \quad (1)$$

where i_0, …, i_n denote the states of the Markov chain on each state. For the case study the Markov chain is presented in Figs. 2, 3, and 4.

3.2 Phase B: Designing the Experiment in the Application Prototype

In the context of management by indicators the most common need is to measure the state of an interesting aspect that is represented by a variable. When talking about indicators the concept of alarm or signal arises together with the semaphore or traffic light as its most common visual representation. The semaphore itself is an indicator that leads to take an action depending on its state (or color). These states and thresholds for indicators are defined by a DS or a subject matter expert depending on his or her experience [1].

3.2.1 Choosing the Analysis Variable (the Indicator)

As it was presented by authors in [1], a finance case study was selected to validate the proposal presented in this paper. The specific application is on the context of finance referred to the determination of the arrears of a clients' portfolio (in collaboration with an appliances' company).

The indicator selected to show a particular case and to corroborate the proposed methodology was Amount of Total Credit (ATC): The total amount of money currently in debt by customers. The data tables presented below show the application of the case study, where all values are presented in Argentine Pesos (denoted by the symbol $ARS); all currency data in $ARS has been preprocessed to move it to a current value taking as a base the inflation index established by the National Bank of Argentina [11]. The definition and monitoring of indicators are part of the control process of the management organization, which is not discussed in this paper.

3.2.2 Defining the Experimentation Scenarios

Two experimentations where defined for this case study as is explained next: First Scenario: It was defined by all available data as a subset of input data to forecast the following scenario. Second Scenario: It was defined by the last period as a subset of input data to forecast the next period (Table 1).

Table 1 Data subset for the indicator *Amount of Total Credit* defined for the second scenario

Date	Value ($ARS)	State	Data subset
01/08/2007	576632.41	ACCEPTABLE	Input data subset for forecasting
…	…	…	
01/05/2010	348582.04	OPTIMAL	
…	…	…	
01/12/2011	925276.6	CRITICAL	
01/01/2012	909961.33	CRITICAL	
01/02/2012	896456.38	ACCEPTABLE	
…	…	…	
01/03/2012	901764.27	ACCEPTABLE	
01/04/2012	918204.12	CRITICAL	Data subset for forecasted value verification
…	…	…	
01/07/2012	904578.51	CRITICAL	

3.3 *Phases C and D: Results Verification*

During the development of the research work presented in [1] it was concluded that the Markov method, according to its theoretical model, contributes for managing organizational knowledge management indicators. The work presented in this paper evaluates the possibility of using an additional instance of the same method by applying the observation of variable order chains.

3.4 *Phase E: Evaluating with Other Methods or with a Modification of the Current Model*

3.4.1 Markov Chains of Order *N*

One of the fundamental premises presented in the introduction of this work is the incorporation of a novel concept based on extending the analysis of the states of a Markov chain to a fixed size window. It is sensed that if a state could predict the next one, using a sequence of consecutive states, it is possible to predict with greater certainty the next state, or in other words, the next sequence of states. The length of this sequence becomes an analysis parameter that needs to be evaluated on each step by the DS.

3.4.2 Concept Explanation

In a Markov chain with a discreet time, the values of X(t) occurs with certain probability. These values can be imagined, one aside the other, as it is shown in Fig. 2; each box represents a possible variable's state, probabilities of passing from one box to another are also considered.

It could be said that in this case the analysis window has only one state. For example all the values of X(t) can be associated with colors Red (R), Green (G) and Yellow (Y) with such transition probabilities that would allow all the states to communicate with each other. A state transition graph could be defined where all the states are linked to the possible values of X(t), the transitions represent the probabilities (Fig. 3).

It is clear that the possible states would be R, G and Y; however if the analysis is extended to two states, the possible states would be these combinations RR, RG, RY, RG, YG, GG, RY, GY, YY. It has to be noticed that this is not a combination of states, it is the probability of having the next value of Xi before a value of Xj, where i, j $\in$ {R; G; Y}. All the possible combinations could be took into account, however their probability would be zero. Each probability represents a probability of change of variable X with value from i to j. It can be observed that all the values are different from zero where the last state match the first. For example RY $\rightarrow$ YG, indicates the probability that having X(t − 1) = R and X(t) = Y the next state would be X(t − 1) = G. It can be simplified as P(X(t + 1) = YG | X(t) = RY). This situation is depicted in Fig. 4.

3.5 *Context for the Application of Markov Chains of Order* N

The application of this concept could help to forecast the next state of an indicator in the database; it could be helpful to predict and estimate tendencies. Tendencies are state's patterns that repeated in a chain. For example a tendency could be Green–Green–Green, this situation indicates a tendency to a Green state. It is also possible to have mixed tendencies such as Green–Red–Green–Red; this situation could indicate a tendency towards a Green–Red state with an order n = 2. Using this concept could be helpful to predict what the expected value of a transition is since there is a tendency.

3.5.1 Testing Results of First-Order Markov Chains

As mentioned before the indicator Amount of Total Credit was chosen to test authors' proposal. In Fig. 5 the results of applying the Markov method in the software prototype developed by authors can be seen. There is a value for the model's order, a value for the indicator, an associated state, and a period of time that is defined by the DS.

The prototype is now in an experimental stage, as part of a degree thesis of the Systems Engineering from the National Central University. It is in development under standards of free software and open source, and it is leaded by the first author of this paper.

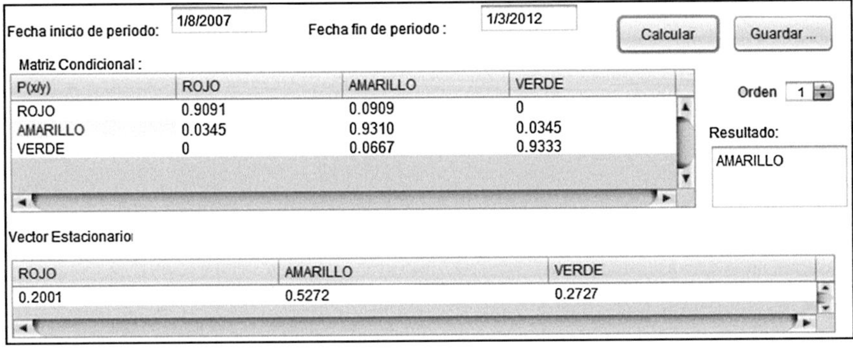

Fig. 5 Probability distribution matrix and stationary vector obtained by computational sampling using the authors' software prototype

Table 2 Transitions Matrix obtained for a Markov chain of order 2

	YY	YG	GG	GY	YR	RR	RY
YY	0.926	0.037	0.000	0.000	0.037	0.000	0.000
YG	0.000	0.000	1.000	0.000	0.000	0.000	0.000
GG	0.000	0.000	0.929	0.071	0.000	0.000	0.000
GY	1.000	0.000	0.000	0.000	0.000	0.000	0.000
YR	0.000	0.000	0.000	0.000	0.000	1.000	0.000
RR	0.000	0.000	0.000	0.000	0.000	0.900	0.100
RY	1.000	0.000	0.000	0.000	0.000	0.000	0.000

Table 3 Stationary vector for a Markov chain of order 2

YY	YG	GG	GY	YR	RR	RY
0.490	0.018	0.256	0.018	0.018	0.181	0.018

Table 4 Stationary vector (summarized) for a Markov chain of order 3

RRR	RRY	RYY
0.81817293	0.09091531	0.09091175

3.5.2 Testing Results of Markov Chains of Order 2

Using the same indicators database from the previous example and the author's software prototype, some testing by grouping two states was made. In Table 2 the transitions matrix that was obtained is shown, next in Table 3, the stationary vector for the Markov Chain of Order 2 is presented. It is important to mention that in order to calculate the stationary vector, all the null rows and columns must be deleted since they represent unreachable states.

3.5.3 Testing Results of Markov Chains of Order *N*

For this particular scenario some testing for grater orders ($n > 2$) has been made and interesting results have emerged. For example for a Markov chain of order 3 the stationary vector (with non-null values) is presented in Table 4; for a Markov Chain of order 4, the state with the highest probability, the resulting state is RRRR.

4 Conclusions

Result analysis for first-order Markov chains. The probability for the Yellow state shown in the stationary vector indicates that in the following time period the indicator has a highest probability to stay in the same state, therefore the indicator must have a value that would be considered as acceptable. The results of the first order belong to the previous work mentioned before [1].

Result analysis for Markov chains of order 2. As it can be seen the stationary vector has been extended to two dimension since the combinations of two ordered states have to be considered. The combination with the highest probability is Yellow–Yellow (YY). It can be observed that the probabilities for Red–Green or Green–Red are null since this occurrence doesn't exist on the indicators database. This fact can be the produced due to the characteristics of the indicator, since the changes from an optimal value to a critical or from a critical value to an optimal in a defaulters' scenario is unlikely. It also can be observed that those states with the highest probability are those pairs with similar values (GG; RR; YY); this can be interpreted as a tendency to maintain the same state instead of changing to a different one.

Overall comparisons. As for the first-order Markov chains as for those of order 2, the states show similar occurrence probabilities. Logically for Markov chains of order 2 the mixed probability of YY is lower than Yellow, however it remains in first place before Green and Yellow remaining in similar proportions. The results for greater orders are more interesting since the combinations RRR with a quite high probability value (0.8181) occur more frequently among the results of the validation data set. The same happens with the combination RRRR that also has a high probability (0.7777) and also is confirmed on those time periods selected for validation.

Forecast verification. Using the remaining dataset for verification, it can be observed that critical states are presented for the next four time periods. Despite there is no coincidence with the state with the highest probability (Yellow or Acceptable), the same value remains as it is suggested in the evolution analysis of the indicators. Results become more interesting if analysis windows are widened.

General conclusions. This paper has tried to reveal information about the state from the indicator stored in a chronologically ordered database. In order to do that, Markov chains have been used to reach the goal of predicting the behavior of the system across the time. The incorporation of a new parameter in the Markov chain, the grouping order of the states, allows to improve the forecasting for new states since the analysis window is widened, this fact was show on the different results from the stationary vectors. The enlargement of the analysis window seems to offer more precise forecasting results when all data is compared; it could be interesting to conduct more testing with a bigger dataset and more variation on the data, or even on different domains, for example, using academic indicators.

On the other hand another of the goals of this work has been achieve by the proposal of the methodological aspects that could lead a DS to conduct an analysis using the phases described in Sect. 2.

The methodology was applied to other case studies: medical services, pharmacy, stocks of spare parts, academic record data; the corresponding analysis could be performed in all of the cases. Due to the extension of this work it is not possible to present the results thereof.

References

1. Illescas, G., Sanchez-Segura, M., Canziani, G.: Comprobación de métodos de pronóstico de indicadores dentro de la gestión del conocimiento organizacional. 3er Congreso Internacional de Mejora de Procesos Software (CIMPS 2014). Centro de Investigación en Matemáticas (CIMAT, Zacatecas A.C.). Octubre 2014, Zacatecas, México (2014)
2. Begleiter, R., El-Yaniv, R., Yona, G.: On prediction using variable order Markov models. J. Artif. Intell. Res. **22**, 385–421 (2004)
3. Cooper, C.: Analytics and big data—reflections from the Teradata Universe conference 2012. Blog post. http://bit.ly/CooperTeradataBlog. Accessed 27 April 2012
4. Kaplan, R., Norton, D.: Using the balanced scorecard as a strategic management system. Harv. Bus. Rev. (1996)
5. Berenson, M., Levine, D.: Estadística Básica en Administración, Conceptos y Aplicaciones. Sexta edición. Prentice Hall Hispanoamericana, S. A. México (1996)
6. Chan, S., Ip, W.: A scorecard—Markov model for new product screening decisions. Department of Industrial and Systems Engineering, The Hong Kong Polytechnic University. Industrial Management & Data Systems, vol. 110, No. 7, pp. 971–992. (c) Emerald Group. (2010)
7. Köppen, V., Allgeier, M., Lenz, H.: Balanced Scorecard Simulator—A Tool for Stochastic Business Figures. Part VI, pp. 457–464. Institute of Information Systems, Free University Berlin, D-14195 Berlin, Germany (2007)
8. Blumenberg, S., Hinz, D., Goethe, J.: Enhancing the prognostic power of it BSC with Bayesian belief networks. University, Frankfurt, Germany. In: Proceedings of the 39th Hawaii International Conference on System Sciences. 0-7695-2507-5/06 (c) IEEE (2006)
9. Prada Alonso, S.: Cadenas de Markov en la investigación del genoma (2013). http://eio.usc.es/pub/mte/index.php/es/trabajos-fin-de-master/finalizados
10. Mascareñas, J.: Procesos estocásticos: introducción. ISSN: 1988–1878. Universidad Complutense de Madrid (2013). http://pendientedemigracion.ucm.es/info/jmas/mon/27.pdf. Accessed 20 June 2015
11. Illescas G., Sanchez-Segura, M., Canziani, G.: Forecasting methods by indicators within the management of organizational knowledge. Revista Ibérica de Sistemas y Tecnologías de Información, pp. 29 a 41. ISSN: 1646-9895-©AISTI 2015 Edición Nº E3, 03/2015. Associação Ibérica de Sistemas e Tecnologias de Informação. Portugal (2015)

Defect Prediction in Software Repositories with Artificial Neural Networks

Ana M. Bautista and Tomas San Feliu

Abstract One of the biggest challenges that software developers face it is to make an accurate defect prediction. Radial basis function neural networks have been used to defect prediction. Software repositories like GitHub repository have been mined to get data about projects and their issues. The number of closed issues could be a useful tool for software managers. In order to predict the number of closed issues in a project, different neural networks have been implemented. The dataset has been segmented by the criterion of project size. The designed neural networks have obtained high correlation coefficients.

Keywords Data mining · Software repositories · Artificial neural networks · Defect prediction

1 Introduction

Developers and users often find issues in software systems and they are encouraged to report them in the available issue trackers that are set up by development teams. There is a variety of issue tracking systems. Popular systems include Bugzilla, Jira, etc. Other development platforms, such as Google Code, GitHub or Freecode, have in-house implementations of issue tracking systems. Developers thus have ample opportunities to use issue-tracking systems in their development and maintenance process [1].

Our goal in this paper is to analyze the relation among the different metrics of each project. Specifically, studying whether size, number of watchers and/or

A.M. Bautista (✉) · T.S. Feliu
Department of Lenguajes Y Sistemas Informaticos E Ingenieria Del Software, Universidad Politécnica de Madrid, 28040 Madrid, Spain
e-mail: am.bautista@alumnos.upm.es

T.S. Feliu
e-mail: tomas.sanfeliu@upm.es

J. Mejia et al. (eds.), *Trends and Applications in Software Engineering*,
Advances in Intelligent Systems and Computing 405,
DOI 10.1007/978-3-319-26285-7_14

number of forks have influence in the number of closed issues of the project and therefore, it is possible to predict this number when you know the others.

Our study exploits data from GitHub, a super-repository of software projects containing millions of projects. GitHub is free for open source projects and implements in-house issues tracking system where users can file issues and tag them into self-defined categories. The issue tracking system is easy to use and it is systematically provided to all projects hosted in GitHub.

The remainder of this paper is organized as follows. In Sect. 2, we introduce the related work about predicting issues using neural networks. In Sect. 3, the proposed approach is shown. We present the dataset and neural network implementations. In Sect. 4, the results achieved are analyzed. In Sect. 5, the challenges and limitations are declared. Finally, in Sect. 6, conclusions and future work are described.

2 Related Work

Different techniques have been used for the purpose of classification/predicting defects. They can be broadly grouped into techniques used for predicting expected number of defects to be found in a software artifact (Prediction) and techniques that are used to predict if a software artifact is likely to contain a defect (Classification) [2]. Among these techniques are machine learning models.

Machine learning models use algorithms based on statistical methods and data mining techniques that can be used for defect classification/predictions. These methods are similar to logistic regression and they use similar input data (independent variables). The key difference is that machine-learning models are based in dynamic learning algorithms that tend to improve their performance, as more data are available. Using code metrics data of projects from NASA IV&V facility Metrics Data Program (MDP), Menzies et al. [3] model based on Naïve Bayes predicted with accuracy of 71 % (pd, probability of detection) and probability of false alarm (pf) of 25 %. Gondra [4] also using NASA project data set (JM1) obtained correct classifications of 72.6 % with ANNs and 87.4 % with SVMs. Using data from 31 projects from industry and using BNNs Fenton et al. [5] obtained an R2 of 0.93 between predicted and actual number of defects.

One of the most important types of machine learning algorithms is artificial neural network. An artificial neural network learning algorithm (ANN), usually called "neural network" (NN), is a learning algorithm that is inspired by the structure and functional aspects of biological neural networks. Computations are structured in terms of an interconnected group of artificial neurons, processing information using a connectionist approach to computation Modern neural networks are non-linear statistical data modeling tools. They are usually used to model complex relationships between inputs and outputs, to find patterns in data, or to capture the statistical structure in an unknown joint probability distribution between observed variables [6, 7].

2.1 *Radial Basis Function Neural Network*

In 1985, Powell proposed multivariate interpolation of Radial Basis Function Method (RBF), which provided a novel and effective mean for the multilayer forward network study. In 1988, Broomhead and Lowe designed the first RBF neural network [8].

RBF differ from MLP in that the overall input-output map is constructed from local contributions of Gaussian axons, require fewer training samples and train faster than MLP. The most widely used method to estimate centers and widths consists on using an unsupervised technique called the k-nearest neighbor rule. The centers of the clusters give the centers of the RBFs and the distance between the clusters provides the width of the Gaussians.

Since then, the RBF neural network has caused wide attention, and a large number of papers about RBF neural network structure, learning algorithm and application in various fields have been published. The application involves many aspects of life now. RBF neural network not only has good generalization ability, but it also can avoid the cumbersome and lengthy like back propagation calculation. Moreover, it also can make learning from 1000 to 10,000 times faster than the normal Back Propagation neural network. Besides, we have decided to use this type of neural network because we have got very good results in previous studies, with NASA93 file of PROMISE, even with only 93 patterns [9].

2.2 *GitHub*

GitHub is one of the most widely used public repositories, with more than 10 million projects, contains a wealth of information about the practice of software development. Further, it has an in-house implementation of an issue tracking system. On the other hand, we have already been working with the download file of more than 145,000 projects written in Python language that are hosted in this repository [10].

3 Proposed Approach

The target of this study is to analyze the relationship between the different metrics existing in a project and the number of issues detected in that project.

3.1 *Dataset*

Firstly, we access to GitHub repository to get project and issues data. We have built two Microsoft PowerShell programs to get these data. Each one of them uses the URL of the project to obtain the information.

Example of main fields in a GitHub project:

"id": 470423,
"name": "py4s",
"full_name": "wwaites/py4s",
"owner":

"login": "wwaites",
"id": 181467,

"private": false,
"description": "Python bindings via C for 4store",
"fork": false,
"size": 1078,
"stargazers_count": 15,
"watchers_count": 15,
"language": "Python",
"has_issues": true,
"has_downloads": true,
"has_wiki": true,
"has_pages": false,
"forks_count": 8,
"open_issues_count": 3,
"forks": 8,
"open_issues": 3,
"watchers": 15,
"default_branch": "master",
"network_count": 8,
"subscribers_count": 1

Github provide two metrics related with the project popularity such as watchers and fork. The "watchers" metric gives an indication of the amount of attention that is given to a project by the developer community. Any user can "watch" a project to receive notifications about events in a project. Similarity forks metrics measure the active involvement of the developer community in the growth of project's code base. Developers often create copies of project code base for continuing the development in parallel, while merging their improvements with the mainline repository. Usually, these metrics can be used for evaluating the project popularity and thus the success of the project.

The project data contain the number of open issues but not the overall of issues. When you retrieve the issues of a project you receive them in inverse order of its creation date. Therefore, the number of issue within the project of the first issue that is recovered is coincident with the total of issues for that project. Using Microsoft Excel, we add this information to the others of the project and we calculate the number of closed issues of the project. The number of closed issue is calculated as the difference between the number of open issues and the overall of issues, previously calculated.

Table 1 Files by size of projects

File	Size	Records
B_P	<300	838
B_M	>300 and <1000	414
B_G	>1000	470

Firstly, we use the 981 projects with issues inside de first group of 5000 projects retrieved (File A).

Then, we retrieved other 5000 projects more and we select the 813 projects with issues inside them. The new file has 1794 patterns (File B).

Later, we decide to group the projects by size to study whether these groups have more homogeneous features. Initially, we remove the projects with wrong size (54 with size equal to zero) and the very big ones (18 with size greater than 100,000). Then, we divide the rest of projects in three files, as shown in Table 1.

Finally, in order to organize the data, these have been separated into groups according to the criteria of the log-normal size distribution. The process is to calculate the natural logarithms of the data, compute the standard deviations and range values on these logarithmic data, and convert back to the antilogarithms. Briefly, we follow these steps:

Step 1. Calculate the natural logarithm of the number of defects.
Step 2. Calculate the average of the logarithmic values
Step 3. Calculate the variance of the logarithmic values around their mean.
Step 4. Using these logarithmic values for the average and standard deviation, we calculate the logarithms of the issues range points.
Step 5. Take the antilogarithms of these to get the size range points.

We calculate the mean (A = 6.24) and the standard deviation (D = 1.6166) of the log-normal distribution of the 1722 projects. We generate nine files, as shown in Table 2.

Table 2 Files as log-normal distribution

File	Size	Log-normal size distribution	Records
P	<103	<(A − D)	184
M	>103 and <2592	>(A − D) and <(A + D)	1248
G	>2592 and <100,000	>(A + D) and <100,000	290
1M	>103 and <513	>(A − D) and <A	872
P1M	>513 and <2592	>A and <(A + D)	376
2M	<513	<A	1056
2MG	>513 and <100,000	>(A + D) and <100,000	666
PMM	<230	<A − (D)/2	706
MMG	>1140 and <100,000	>A + (D)/2 and <100,000	436

3.2 Neural Network Implementation

We select the fields of project to train the neural network. We consider the size, the number of forks, the number of watchers, the number of open issues and the calculated fields, the number of closed issues and the overall number of issues as the best ones to train the neural network and to extract information.

We use radial basis function with learning rule momentum, a Euclidean metric and tanhAxon as function transfer. The net uses a competitive rule with full conscience in the hidden layer and one output layer with the Tanh function, all the learning process has been performed with the momentum algorithm. The TanhAxon applies a bias and tanh function to each neuron in the layer. Such nonlinear elements provide a network with the ability to make soft decisions.

To build the neural network we have use the NeuroSolutions tool. This is a useful tool because it allows generating and testing different models in a few minutes what it is essential during training process. The input to the network is a text file whose fields are separated by commas. The number of the patterns in the data set cases 80 % was used as training, and 20 % for testing.

We have built different neural networks and we have fit to achieve the best possible results. We explain now each one.

- Network 1 has two input neurons: size and fork. The main goal is predicting the number of issues, in a project, whose state is closed,
- Network 2 has three input neurons: size, forks and total number of issues, one output neuron: number of issues whose state is closed and one hidden layer, with ten clusters. The patterns are presented to the network 10,000 times or 50,000 times.
- Network 3 we change, in network 2, total number of issues by number of issues whose state is open, in the input layer.
- Network 4 we change, in network 2, forks by watchers, in the input layer.
- Network 5 we change, in network 6, total number of issues by number of issues whose state is open, in the input layer.

4 Results

Network 1. With two input neurons: size and forks. This network predicts the number of issues, in a project, whose state is closed.

The results improve when we submit the File B_P, B_M and B_G to the Network 1. We get the best result for File B_G, with Mean Squared Error (MSE) equal to 0.002, correlation coefficient (r) equal to 0.90 and 100 % of true positives and negatives. Checking the good results to split the original file by sizes, we decide to create nine files by statistic analysis, according to Table 2.

Network 2. With three input neurons: size, forks and total number of issues, one output neuron: number of issues whose state is closed.

The results with the nine files and the Network 2 are very good, as shown in Table 3, because MSE is the thousandths for eight files, and r-correlation coefficient is greater than 0.80. The percentage of true positives and negatives is always upper than 85 in each and every files.

Network 3 has three input neurons: size, forks and total number of issues. And it has one output neuron: number of issues whose state is closed.

The best results by applying the nine files and the Network 3 are shown in Table 4.

Generally, the results are worse than the execution of Network 2. The difference between the two networks was the input neuron total number of issues and number of open issues. Therefore, the number of open issues does not improve with respect the total number of issues.

Network 4 has three input neurons: size, watchers and total number of issues. The output of network 4 is the number of issues whose state is closed.

The results with the nine files and the Network 4 are very good, as shown in Table 5, because MSE is the thousandths for all files, and r-correlation is greater than 0.85. The percentage of true positives is upper than 99 in each and every file. The percentage of true negatives is 100 in all but two files.

Network 5 has three input neurons: size, watchers and issues open. The output neuron is the number of issues whose state is marked as closed.

Table 3 Results by applying neural network 2

Files	MSE	r	True positives (%)	True negatives (%)
P	0.010	0.81	100	100
M	0.002	0.92	99.67	100
G	0.001	0.96	100	100
1 M	0.008	0.89	99.41	93.33
P1 M	0.006	0.90	99.51	86.66
2 M	0.003	0.96	100	100
2 MG	0.001	0.93	100	100
PMM	0.003	0.85	99.85	100
MMG	0.001	0.93	100	100

Table 4 Results by applying neural network 3

Files	MSE	r	True positives (%)	True negatives (%)
P	0.020	0.70	100	100
M	0.009	0.68	99.91	20
G	0.020	0.59	99.65	100
2M	0.010	0.78	99.73	80
2MG	0.003	0.79	99.84	50
MMG	0.002	0.89	100	100

Table 5 Results by applying neural network 4

Files	MSE	r	True positives (%)	True negatives (%)
P	0.001	0.98	99.45	100
M	0.003	0.91	99.67	100
G	0.002	0.94	100	100
1 M	0.009	0.87	99.53	60
P1 M	0.007	0.88	99.61	80
2 M	0.002	0.91	100	100
2MG	0.001	0.94	100	100
PMM	0.003	0.86	99.85	100
MMG	0.001	0.94	100	100

Table 6 Results by applying neural network 5

Files	MSE	r	True positives (%)	True negatives
P	0.010	0.72	100	100
M	0.006	0.80	100	80
G	0.003	0.90	100	100
1 M	0.020	0.65	99.88	26.66
P1 M	0.020	0.59	100	6.66
2 M	0.010	0.81	99.73	20
2MG	0.002	0.87	100	100
MMG	0.003	0.87	100	100

The results by applying the nine files and the Network 5 are shown in Table 6. Generally, the results are worse than when we apply the Network 4, specially, for small projects. We selected Network 4 due to the improved performance. As in the previous case, the behavior of total number of issues has been better than number of open issues.

5 Challenges and Limitations

Using GitHub to treat large volumes of information it is slow. It is possible to retrieve only 5000 projects per hour. In this paper we have limited our study to a snapshot of 1722 projects written in Python language. Our sample set of projects may not represent the universe of all real world projects. Our dataset cannot be use to conclude causation instead it can only show correlation. Our findings must be carefully considered within the context of open source projects.

Moreover, the database of GitHub is constantly changing. For instance, one user can make one fork or a new watcher can attach to a project. This means that a minute or a second later, a project has one fork more or one watcher more that when we retrieved the data of the mentioned project.

6 Conclusions and Future Work

Software repositories have traditionally been used for archival purposes. Issue reports are important artifacts in software development. Unfortunately, the process of acquiring such information in a convenient format is challenging, since such repositories are mainly designed as record keeping repositories. In this paper we have checked neural networks are useful tools to mine in software repositories and to predict the number of issues that will be closed in a project.

The neural networks have obtained better results with big projects. These are good news because defect prediction is more necessary for big projects. Big projects involve more resources to manage issues.

In this paper, we have obtained that size; forks and total number of issues help us to predict number of issues that will be closed. We can also use number of watchers to estimate number of closed issues.

As we now have a dataset with information about the timing of issues, the study could be extended to other models that describe the issues arrival patterns. Other potential models to study are those related to predicting when an issue may be closed.

Acknowledgments This work is sponsored by everis Aeroespacial y Defensa, and the Universidad Politecnica de Madrid through the Research Chair of Software Process Improvement for Spain and Latin American Region.

References

1. Bissyande, T.F., et al.: Got issues? Who cares about it? A large scale investigation of issue trackers from github. In: 2013 IEEE 24th International Symposium on Software Reliability Engineering (ISSRE), pp. 188–197. IEEE (2013)
2. Rana, R.: Software defect prediction techniques in automotive domain: evaluation, selection and adoption. Doctoral dissertation, University of Gothenburg (2015)
3. Menzies, T., Greenwald, J., Frank, A.: Data mining static code attributes to learn defect predictors. IEEE Trans. Softw. Eng. **33**(1), 2–13 (2007)
4. Gondra, I.: Applying machine learning to software fault-proneness prediction. J. Syst. Softw. **81**(2), 186–195 (2008)
5. Fenton, N., Neil, M., Marsh, W., Hearty, P., Radliński, Ł., Krause, P.: On the effectiveness of early life cycle defect prediction with Bayesian Nets. Empir. Softw. Eng. **13**(5), 499–537 (2008)
6. Gareth, J., Witten, D., Hastie, T., Tibshirani, R.: An Introduction to Statistical Learning. Springer, New York (2013)
7. Mohri, M., Rostamizadeh, A., Talwalkar, A.: Foundations of Machine Learning. MIT Press, Cambridge (2012)
8. Broomhead, D.S., Lowe, D. Radial basis functions, multi-variable functional interpolation and adaptative networks. (No. RSRE-MEMO 4148) Royal Signals and radar Establishment Malvern, UK(1988)

9. Bautista, A.M., Castellanos, A. San Feliu, T.: Software effort estimation using radial basis function neural networks. Inf. Theor. Appl. 319 (2014)
10. Bautista, A.M., San Feliu, T: A process to mining issues of software repositories. In: 2015 10th Iberian Conference on Information Systems and Technologies (CISTI). IEEE (2015)

Part III
Software Systems, Applications and Tools

Reverse Engineering Process for the Generation of UML Diagrams in Web Services Composition

Isaac Machorro-Cano, Yaralitset López-Ramírez, Mónica Guadalupe Segura-Ozuna, Giner Alor-Hernández and Lisbeth Rodríguez-Mazahua

Abstract Reverse engineering is a process that allows us to optimize and reuse code and applications based on a complete analysis of such code. This permits us to generate a new idea that it is based on an already existent one and it creates or improves the tool in which it is applied. In addition, UML activity diagrams are graphic representations of processes that facilitate the way in which we perceive and understand the order of execution of individual processes. It also provides visual representation of the primary subsystems that define the entire process. This work presents the application of reverse engineering to a transformation mechanism in WSCDL, which allows the generation of UML activity diagrams from WSCDL documents, and includes a case study to exemplify the functionality of this mechanism.

Keywords Reverse engineering · Transformation mechanism · WSCDL · Activity diagrams

I. Machorro-Cano (✉) · M.G. Segura-Ozuna
Universidad Del Papaloapan (UNPA), Tuxtepec, OAX, Mexico
e-mail: imachorro@unpa.edu.mx

M.G. Segura-Ozuna
e-mail: yarita26ra@gmail.com

Y. López-Ramírez
Instituto de Estudios Superiores de Oaxaca A.C.(IESO), Oaxaca, OAX, Mexico
e-mail: msegura@unpa.edu.mx

G. Alor-Hernández · L. Rodríguez-Mazahua
Instituto Tecnológico de Orizaba (ITO), Orizaba, VER, Mexico
e-mail: galor@itorizaba.edu.mx

L. Rodríguez-Mazahua
e-mail: lrodriguez@itorizaba.edu.mx

J. Mejia et al. (eds.), *Trends and Applications in Software Engineering*,
Advances in Intelligent Systems and Computing 405,
DOI 10.1007/978-3-319-26285-7_15

1 Introduction

Reverse engineering is a method that allows us to extend the useful life of existing programs, also known as legacy software. Since reverse engineering leads to the restructuring of existing code, it is considered the opposite of code generation. This process involves examining and analyzing, legacy applications with the intention of extracting entire code sequences that will be incorporated into future development. The advantage of using reverse engineering in this manner is a reduction in development time for both the creation of new programs and the maintenance/upgrade of existing ones [1]. There are two types of reverse engineering: (1) Source Code-based: Where the source code is available, but full knowledge of its business process value is either unknown or poorly documented; (2) Executable program-based: Where the source code is not available, so the external measures must be employed in an effort to reveal the underlying structure.

A major objective, when conducting both types of reverse engineering, is to clearly define the relationship between program function and workflow requirements. An excellent tool for defining these relationships is the Unified Modeling Language (UML). This is especially true for a sub function of UML know as Activity Diagrams (AD)s. ADs, which unambiguously demonstrate the interaction of various sub-functions in the course of performing a business process, not only illustrate the order of execution of various activities, but account for the handling of concurrent processes, i.e., the sequencing rules that have to be followed [2]. Due to the hierarchical nature of UML, system diagrams that are implemented using ADs can be viewed from either a high-level perspective or with step-by step detail. These diagrams are especially useful because they focus attention on how each process is sequenced and the moment of its synchronization with other activities. The key symbol of this type of diagram is the activity; this represents certain actions executed either by user interaction or by a background process. Furthermore, ADs are used to represent the logic of complex operations, business roles, use cases and concurrent processes [3]. In this endeavor, the application of reverse engineering provides a transformation mechanism that is typically represented using WSCDL (Web Services Choreography Description Language). WSCDL also supports the direct translation of ADs into java classes. This suggests that WSCDL is an optimal tool for both Software-based and Execution-based reverse engineering.

We also present a case study of the search for supplier that offers the lowest price available for an electronic component that provides the required functionality. This work is structured as follows: Sect. 2, which presents the description of the transformation mechanism that was employed in WSCDL; Sect. 3, which describes related works; Sect. 4, which explains the application of reverse engineering to the proposed transformation mechanism; Sect. 5, which presents the case study of the "Search of the provider which offers the lowest price of an electronic component"; Finally, Sect. 6, which offers conclusions and a plan for future work.

2 Mechanism of Transformation in WSCDL

Currently, new complex business processes are modeled using the technologies of Web Services (WS) and UML, such processes need mechanisms to efficiently interoperate at a business level. WS is a recently developed platform that helps maximize interoperability in distributed applications by optimally integrating sub-services to generate new business processes [4]. In this context, there are two perspectives for the composition of WS, the orchestration and the choreography. Two composition languages for the realization of the orchestration and the choreography of WS are: WSCDL and BPEL (Business Process Execution Language for Web Services) [5].

We chose WSCDL over BPEL because WSCDL highlights the global exchange of messages among all participants, whereas BPEL provides this information from the point of view of a single participant. Furthermore, WSCDL offers rules that each participant uses to estimate the state of the choreography and to deduce which would be the next exchange of messages while BPEL specifies rules that are executed to deduce the subsequent activities to execute, once the rule is calculated, the time of execution of the orchestration executes the corresponding activities. Also, WSCDL offers a model for the specification of the better understanding of the exchange of messages and BPEL does not provide this support. [6]. Figure 1 presents a diagram of the transformation mechanism in WSCDL. Reverse engineering was applied to this mechanism. The arrows that progress from left to right represent the flow of the WSCDL transformation; whereas those arrows progressing from right to left represent reverse engineering.

In the following section, we present some related works with this proposal. UML diagrams, WSCDL and other composition languages, WS and the programming language Java are used in these works and other aspects.

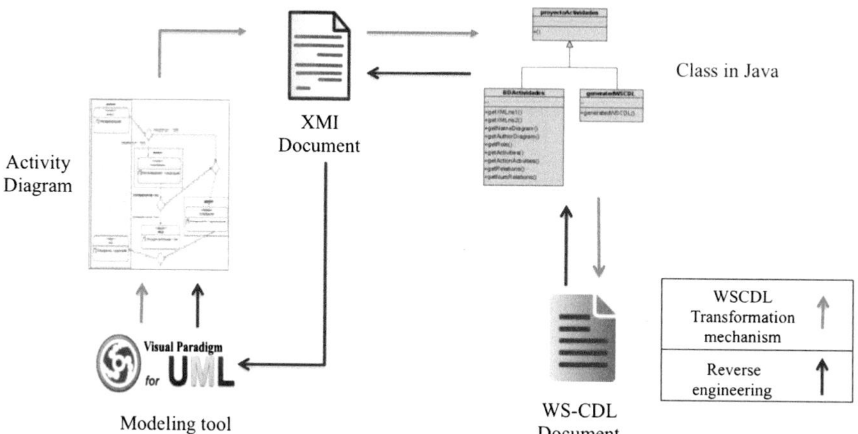

Fig. 1 Transformation mechanism in WSCDL—reverse engineering

3 Related Works

A method which simplified the tedious and complex tasks during the modeling of the operations of the specification of conceptual schemas (CSs) that define the system behavior was proposed in [7]. This method automatically generates a set of basic operations that complement the static aspects of the CS. On the other hand, a systematic study of the empiric studies in relation to the use of UML diagrams for the maintenance of the source code and of the UML diagrams themselves was presented in [8]. In such work it was indicated that it exists a need of more experiments and case studies in real contexts to show the utility of the UML diagrams in the maintenance of the code. Likewise, the evidence about the utility of the UML diagrams during the maintenance of the software was presented in [9]. Also, it was appreciated a tendency in favor of using the redirect of the design of UML diagrams with relation to use the inverse engineering of UML diagrams, because of the effectiveness and efficiency presented by them.

On the other hand, software sizing at a finer level of granularity by taking into account the structural aspect of a sequence diagram (SD), in order to quantify its structural size was explored in [10]. The COSMIC method was used to achieve this objective. The COSMIC method takes into account the data movements in a sequence diagram. Moreover [11] presented the combination of the software testing with "Extenics" for dealing with contradictory problems with formalized methods and transformation. This work proposed an automatic approach that generated test cases from ADs based on the extension theory. Also, a revision of SDs was presented in [12], where STAIRS was used, a formal language to model interactive UML diagrams and to produce models, as an alternative to formalize the diagrams of interaction in UML, specifically the SDs.

Furthermore, a structured approach to automatically annotate a Web page with Rich Snippets RDFa tags was presented in [13]. This work exploited a data reverse engineering method, combined with several heuristics and a named entity recognition technique. In addition, it evaluated the accuracy of the approach on real E-commerce Web sites. In [14] it is demonstrated a mechanism that detected carefully constructed virtual environments by focusing on the stochastic of system call timings. This work also presented a statistical technique for detecting emulated environments to successfully identify the attacks of the hackers, which used the reverse engineering as a powerful tool. On the other hand, an instrument for the characterization of reverse engineering tools was proposed in [15]. Also, it was defined a new characterization structure based on two criteria, the structural aspect and the common properties between them. The instrument was validated in situations that proved his usefulness for classifying and evaluating such tools. The proposal allowed the independent and joint evaluation of each of the components that form reverse engineering tools. Moreover, in [16] it was presented the application of database (DB) reverse engineering through a chain of model-to-model transformations based on a set of meta-models. In addition, this work presented a case study illustrating an approach to DB reverse engineering.

In the following section, we present the application of the reverse engineering to transformation mechanism in WSCDL, where describe the five phases of the used methodology, the developed classes in Java and the generation of the ADs.

4 Application of Reverse Engineering

The transformation process described in this proposal is based on Source-Code reverse engineering and relied on the transformation mechanism provided in WSCDL, as described in Sect. 2. By adhering to Dynamic Systems Development Method (DSDM), we enjoyed the benefits of an iterative and incremental development cycle that offers the ability to easily and reliably reverse all changes made to the original source code during the development. This methodology also integrates tests throughout the life cycle of the project that allow active participation by all interested parties [17]. These are five phases of the DSDM methodology:

Phase 1: Feasibility Study—The importance of applying reverse engineering to the transformation mechanism is identified in this phase with the objective of validating it and generating an ADs from a WSCDL document.

Phase 2: Business Study—A case study is provided below that describes the search for a provider that offers the lowest price on an electronic component capable of delivering the necessary functionality and reliability. The process of this mechanism was studied and is presented in the Table 1. Also, the structure of the corresponding WSCDL document and its relation to an XMI document was considered.

Phase 3: Iteration of the functional model—The following steps were performed in this phase: (a) Revision of the prototype—The analysis of the process and the phases which are is responsible for the transformation of an ADs to a WSCDL document was completed, the employed classes and functions in Java were analyzed. (b) Identification of the functional prototype—The key parts of the transformation mechanism was identified, a case study was selected for verification, and the reusable code submitted for transformation. (c) Plan of action—Reverse engineering was implemented using DSDM methodology as well as the horseshoe method, which supports the isolation and adaptation of reusable code.

Phase 4: Iteration of the design and construction—In this phase we completed the process of generating an XMI document from a WSCDL document where the following actions were performed: (a) Identification of the prototype of the design—The ADs which models the case study used in the transformation mechanism in WSCDL was analyzed, and the different activities that are necessary for its modeling were identified. From the creation of the WSCDL document through the mechanism of transformation, an analysis of each tag was made with the purpose of identifying its relation to a corresponding tag inside an XMI file, and this XMI file was used to generate ADs; (b) Agree on a plan—The tags of the WSCDL file that maintain the relationship with the tags of the XMI file are the following: *node*—This tag is generated from the tag in WSCDL *RoleType*, which represents the set of roles

Table 1 Relationship between WSCDL and XMI

WSCDL code		XMI Document
<package>		<uml:Diagram>
roleType>		<ownedMember>
<participantType>		<uml:Model>
<relashionsipType>	<choreography> <variable Definitions>	
<exchange>	<sequence>	<include>
<send>	<interaction>	addition (Source)
response request	<participate> <relationship Type>	
<recibe>	<exchange> <send>	xmi:id (Destination)
response request	response request	
<exchange>	<recibe>	<extends>
<send>	response request	extended case (Source)
response request	</exchange>	
<recibe>	. .	xmi:id (Destination)
response request	. <exchange>	
<exchange>	<send>	<uml: Dependency>
<send>	response request	client (Source)
response request	<recibe>	
<recibe>	response request	supplier (Destination)
response request	</exchange> </choreography>	
<informationType>		<ownedMember>
<token> <tokenLocator		

inside the diagram; *containedNode*—This tag comes from the tag *informationType* of the WSCDL code, with its attribute name, this tag mentions the name of all the variables inside the diagram and its pertinence in the corresponding node; *edge*—This tag corresponds to the relationship tag inside the WSCDL choreography and uses its ID to derive its attribute source and the target of each variable from the relationship tag; c) Creation of the design prototype—Once the tags and the relationship between WSCDL and XMI had been established, we defined four Java classes to begin implementation of the reverse engineering transformation in WS-CDL, first, generating its corresponding XMI document and then, with the help of the modeling tool, we generated the corresponding ADs. In the programming of

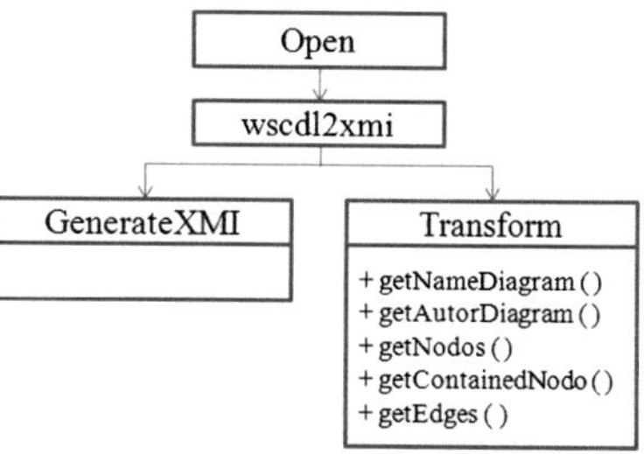

Fig. 2 Class in Java

the Java classes, the libraries DOM (Document Object Model) and SAX (Simple API for XML) were also used. The API DOM describes how a document XML is organized and it allowed us to know how the tags are inserted. The API SAX, aids in the proper interpretation of an XML file and detects when an element of the document begins and ends, it also confirmed that our XML files were well formed. Figure 2 presents the four classes that were developed in Java where:

(1) The **class "Open"** is the main class and from which a Frame is executed, where the corresponding WSCDL is selected and its contents are displayed. From this display we can initiate the generation of its corresponding XMI document. This action invokes the method execute () from the class "wscdl2xmi", and the obtained tags are sent to this latter class.
(2) The **class "wscdl2xmi"** invokes the sub-classes "Transform" and "GenerateXMI" which are responsible for generating the XMI document.
(3) The **class "Transform"** is responsible for obtaining the corresponding tags from WSCDL and transferring them to the XMI document. It contains the following methods: getNameDiagram (): which gets the name of the diagram; getAuthorDiagram (): which obtains the name of the author of such diagram;
(4) The **class "GenerateXMI"** is responsible for building the XMI document and it only has the main method generateXMI (), which, which the extracted information in the class Transform, proceeds to form the final XMI document; d) Revision of the design prototype—We completed the revision of the XMI document generated by the classes in Java by using the tools CASE Visual Paradigm for UML Enterprise Edition, ArgoUML, Poseidon and Eclipse, in order to validate the document and to verify its performance.

Phase 5: Implementation—The implementation and testing necessary to generate the corresponding UML diagram from the WSCDL document were completed during this phase. Once the XMI document had been obtained by the developed classes in Java, the ADs was generated from the XMI document with the modeling tool Visual Paradigm for UML Enterprise Edition. As it is shown in the Table 2, we can see fragments of WSCDL code in relation to the ADs of the proposed use case. In the next section, we present a case study of the search of a provider that offers the lowest price of an electronic component. The used transformation mechanism in WSCDL is based in such case study and it is used for the validation and application of the reverse engineering.

Table 2 Relationship between WSCDL and UML activity diagram

WSCDL Code	Activity diagram
<informationType name="**EnterproductnameandnumberType**" type="**EnterproductnameandnumberMsg**"><description type="**documentation**">Enterproductnameandnumber Message</description></informationType>	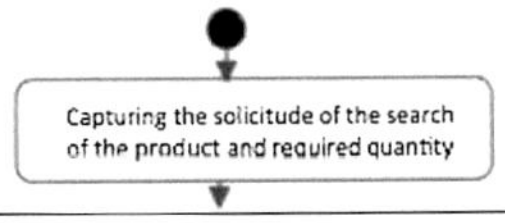 Capturing the solicitude of the search of the product and required quantity
<tokenLocator informationType="**Indicatenumberofcomponentsin existenceType**" tokenName="**tns:id**" query=" "><description type="**documentation**">Identity for Indicatenumberofcomponentsin existence</description></tokenLocator>	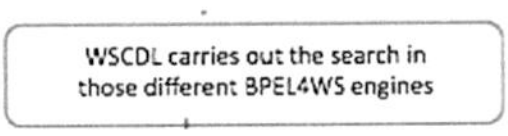 WSCDL carries out the search in those different BPEL4WS engines
<roleType name="**WSDLRole**"><description type="**documentation**">Role for WSDL</description><behavior name="**WSDLBehavior**" interface="**WSDLBehaviorInterface**"><description type="**documentation**">Behavior for WSDL</description></behavior></roleType>	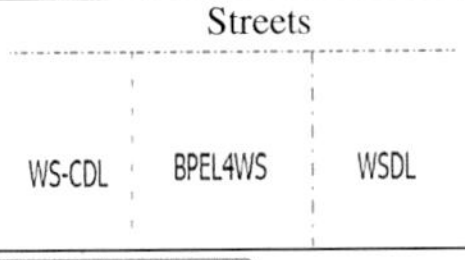 Streets WS-CDL BPEL4WS WSDL
<relationshipType name="**WS-CDL2BPEL4WS**"><description type="**documentation**">WS-CDL BPEL4WS Relationship</description><roleType typeRef="**tns:WS-CDLRole**"/><roleType typeRef="**tns:BPEL4WSRole**"/></relationshipType>	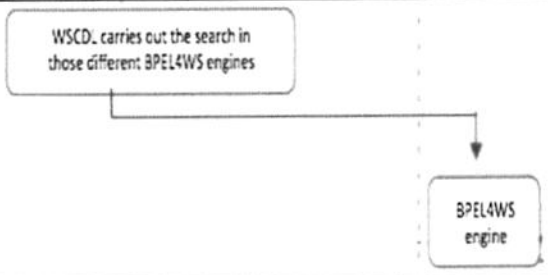 WSCDL carries out the search in those different BPEL4WS engines BPEL4WS engine
<participantType name="**WSDL**"><description type="**documentation**">WSDL Participant</description><roleType typeRef="**tns:WSDLRole**"/></participantType>	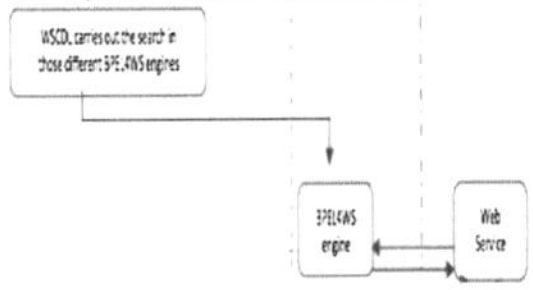 WSCDL carries out the search in those different BPEL4WS engines BPEL4WS engine Web Service
<channelType name="**WS-CDL2BPEL4WSChannel**"><description type="**documentation**">WS-CDL to BPEL4WS Channel Type</description><roleType typeRef="**tns:BPEL4WSRole**"/><reference><token name="**tns:URI**"/></reference><identity><token name="**tns:id**"/></identity></channelType>	 BPEL4WS BPEL4WS engine
<choreography name="**search-productChoreography**" root="**true**">	
<exchange name=" **Generatelistofcomponents** " informationType="**tns: GeneratelistofcomponentsType**" action="**request**"><description type="**documentation**">Request Message Exchange</description><send variable="**cdl:getVariable ('Generatelistofcomponents','','')**"/><receive variable="**cdl:getVariable('Generatelistofcomponents','','')**"/></exchange>	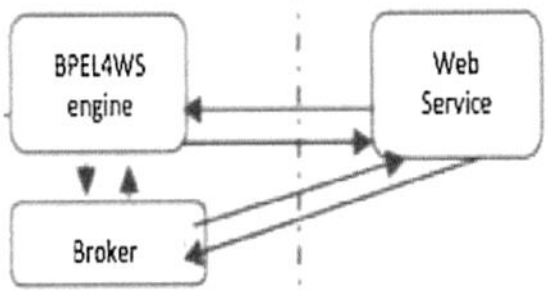 BPEL4WS engine Web Service Broker
<variable name="**GenerateResults**" informationType="**tns:GeneraResultadosType**" roleTypes="**tns:WS-CDLRole tns:WS-CDLRole**"><description type="**documentation**">Request Message</description></variable>	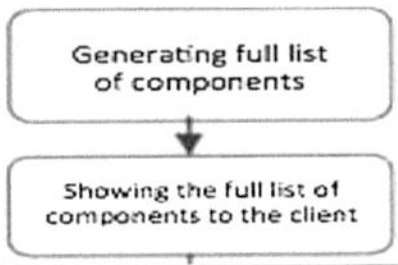 Generating full list of components Showing the full list of components to the client
</choreography>	

5 Case Study: "The Search Provider that Offers the Lowest Price of an Electronic Component"

The most common method of studying business behavior is the case study. A case study describes a situation, lists the steps that were taken to achieve a specific objective and then reflects on the observable outcome. In this regard case studies can be considered descriptive research.

In order to show the validation of the application of reverse engineering to the transformation mechanism in WSCDL, imagine the following scenario: a customer wants to find electronic devices, and the customer decides to use a BPEL broker service in order to find the best device prices. As there are various kinds of BPEL brokering services, this scenario examines how a customer might be able to find the supplier that offers the best combination of price, performance and reliability.

Figure 3 shows the entities involved in this e-commerce scenario, where a costumer accesses a Web site only, which is in communication with the WSCDL engine, which also communicates with the various BPEL engines, which in turn collaborate with intermediaries and other Web services. It is worth mentioning that this scenario is relevant, as it was used in the transformation mechanism described in Sect. 2 and as well as in this study to validate the application of reverse engineering to the transformation mechanism in WSCDL.

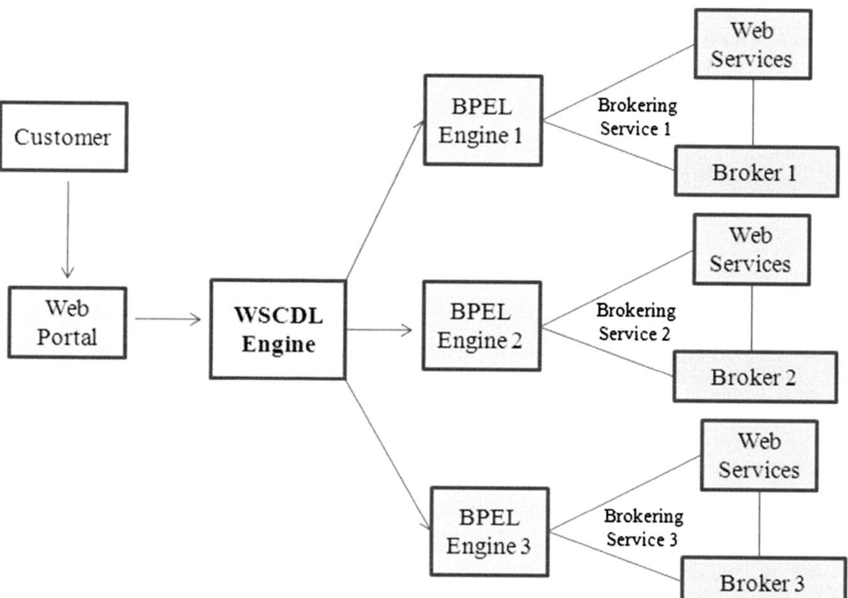

Fig. 3 Case study of the purchase of an electronic component

The conclusions obtained by the application of reverse engineering to a transformation mechanism in WSCDL are presented in the following section. Section 6 also describes future work and gives acknowledgements.

6 Conclusions and Future Work

As stated earlier, reverse engineering is a method that allows us to extend the useful life of existing programs, also known as legacy software. Since reverse engineering leads to the restructuring of existing code, it is considered the opposite of code generation because it consists of taking source code examining and analyzing it and ultimately converting it into entities that can be used for future development. On the other hand, in the execution of a Web service, the activities between services are represented by means of an ADs and the resulting choreography of those activities is defined in WSCDL. In this work we presented the application of reverse engineering to such a mechanism based on the case study of the search for an optimal supplier. This project required the generation of an ADs from a WSCDL document. It was also necessary to develop JAVA classes by reading the WSCDL document and to create the corresponding XMI document that could be analyzed using various modeling tools such as the Visual Paradigm for UML Enterprise Edition.

Finally, the future work consists of applying reverse engineering to other transformation mechanisms in WSCDL that use UML sequences and use case diagrams as the continuation of this work.

Acknowledgments The Authors wish to thank the following institutions: Program for the Teaching Professional Development (PRODEP), The National Council of Science and Technology (CONACYT) and to the National Technologic of México, for the support granted for the realization of the research.

References

1. Park, S.C., Ko, M., Chang, M.: A reverse engineering approach to generate a virtual plant model for PLC simulation. Int. J. Adv. Manuf. Technol. **69**, 2459–2469 (2013). Springer
2. El-Attar, M.: From misuse cases to mal-activity diagrams: bridging the gap between functional security analysis and design: Softw. Syst. Model. **13**, 173–190 (2012). Springer
3. Ranjitha Kumari, S., Panthi, V., Prasad Mohapatra, D., Kumar Behera, P.: Prioritizing test scenarios from UML communication and activity diagrams. Innov. Syst. Softw. Eng. **10**, 165–180 (2014). Springer
4. Hofstede, A., Mecella, M., Sardina S.: Special issue on: knowledge-intensive business processes. J. Data Semant. **4** (2015). Springer
5. Nagamouttou, D., Egambaram, L., Krishnan, M., Narasingam, P.: A verification strategy for web services composition using enhanced stacked automata model. SpringerPlus **4**, 1–13. (2015). Springer

6. Alor Hernández, G., Machorro Cano, I., Gómez, J. M., Cruz Ahuactzi, J., Posada Gómez, R., Mencke, M., Juarez Martinez, U.: Mapping UML diagrams for generating WSCDL code. In: Third International Conference on Digital Society (ICDS), pp. 229–234. IEEE (2009)
7. Albert, M., Cabot J., Gómez, C., Palechano, V.: Generating operation specifications from UML class diagrams: a model transformation approach. Data Knowl. Eng. **70**, 365–389 (2011). Elsevier
8. Fernández Sáez, AM., Genero, M., Chaudron Michel, R.V.: Empirical studies concerning the maintenance of UML diagrams and their use in the maintenance of code: a systematic mapping study. Inf. Softw. Technol. **55**, 1119–1142 (2013). Elsevier
9. Fernández Sáez, AM., Genero, M., Chaudron Michel, R.V., Caivano, D., Ramos, I.: Are forward designed or reverse-engineered UML diagrams more helpful for code maintenance?: A family of experiments. Inf. Softw. Technol. **57**, 644–663 (2014). Elsevier
10. Sellami, A., Hakim, H., Abran, A., Ben-Abdallah, H.: A measurement method for sizing the structure of UML sequence diagrams. Inf. Softw. Technol. **59**, 222–232 (2015). Elsevier
11. Li, L., Li, X., He, T., Xiong, J.: Extenics-based test case generation for UML activity diagram. Proc. Comput. Sci. **17**, 1186–1193 (2013). Elsevier
12. Vidal, C.L., López, L.P., Rivero, S.E., Meza, R.O.: Extension of UML sequence diagrams for aspect-oriented modeling. Inf. Tecnol. **24**(5), 3–12 (2013)
13. De Virgilio, R., Frasincar, F., Hop, W., Lachner, S.: A reverse engineering approach for automatic annotation of Web pages. Multimed. Tools Appl. **64**, 119–140 (2011). Springer
14. Parveen, T., Tilley, S., Allen, W., Marin, G., Ford, R.: Detecting emulated environments. Int. J. Softw. Eng. Knowl. Eng. **22**(7), 927–944 (2012). World Scientific
15. Monroy, M.E., Arciniegas, J.L., Rodríguez, J.C.: Characterization of reverse engineering tools. Inf. Tecnol. **23**(6), 31–42 (2012)
16. Ristic, S., Aleksic, S., Celikovic, M., Dimitrieski, V., Lukovicm I.: Database reverse engineering based on meta-models. Central Eur. J. Comput. Sci. **4**(3), 150–159 (2014). Springer
17. Bustard, D., Wilkie, G., Greer, D.: Towards optimal software engineering: learning from agile practice. Innov. Syst. Softw. Eng. **9**, 191–200 (2013). Springer

Evaluating and Comparing Perceptions Between Undergraduate Students and Practitioners in Controlled Experiments for Requirements Prioritization

José Antonio Pow-Sang

Abstract The Use Case Precedence Diagram (UCPD) is a technique that addresses the problem of determining the construction sequence or prioritization of a software product from the developer's perspective in software development projects. This paper presents a perceptions evaluation of the UCPD based on the Method Evaluation Model (MEM), where the intention to use a method is determined by the users' perceptions. Perceptions were collected using a questionnaire which was applied immediately after the controlled experiments were carried out. Those controlled experiments were replicated twice with undergraduate students, and twice with practitioners. The results show that the intentions to use UCPD exist in undergraduate students and practitioners with at least 2 years of experience in the industry, but the relationships defined by the MEM are best confirmed with the results obtained with practitioners.

Keywords UCPD · Requirements precedence · Requirements prioritization · Software engineering experimentation · Method evaluation model · Method adoption model

1 Introduction

The Use Case Precedence Diagram (UCPD) [1] is a technique based on use cases [2] and its objective is to determine software construction sequences taking into consideration the developer's perspective in terms of ease of construction to define software requirements priorities.

J.A. Pow-Sang (✉)
Departamento de Ingeniería, Pontificia Universidad Católica del Perú,
Av. Universitaria 1801, San Miguel, Lima 32, Peru
e-mail: japowsang@pucp.edu.pe

J. Mejia et al. (eds.), *Trends and Applications in Software Engineering*,
Advances in Intelligent Systems and Computing 405,
DOI 10.1007/978-3-319-26285-7_16

According to UML, the relations that can exist between use cases are: include, extend, and generalization. In addition to the standard, it was proposed the inclusion of a new relation: antecede; and all these relations will be shown in the use case precedence diagram (UCPD). The idea of this diagram was taken from Doug Rosenberg [3], who proposed the use of a similar diagram, specifying the relations "precedes" and "invoke" to determine user requirements (Fig. 1 shows an example of a UCPD). In previous papers [1, 4], the precedence relation was named "precede", but in order to differentiate with Rosenberg's approach this relation was renamed as "antecede", as it is shown in Figs. 1, 2, and 3.

In order to obtain precedence relations between use cases, the following rules must be considered:

Rule 1: A use case U1 precedes another use case U2 if there is a precondition that corresponds to the execution of a scenario in U1 that must be fulfilled before executing a scenario of U2.

For instance, to execute a scenario from the "Make Reservation" use case, the actor must have been validated by the system (i.e. execute a login). Hence, the "Login" use case precedes the "Make Reservation" use case. This is shown in Fig. 2.

Rule 2: A use case U1 precedes another use case U2 if a U2 needs information that is registered by U1. For instance, to perform the payment for a reservation, this reservation must have been made. Having two use cases "Make Reservation" and "Pay Reservation", the former precedes the later. This is shown in Fig. 3.

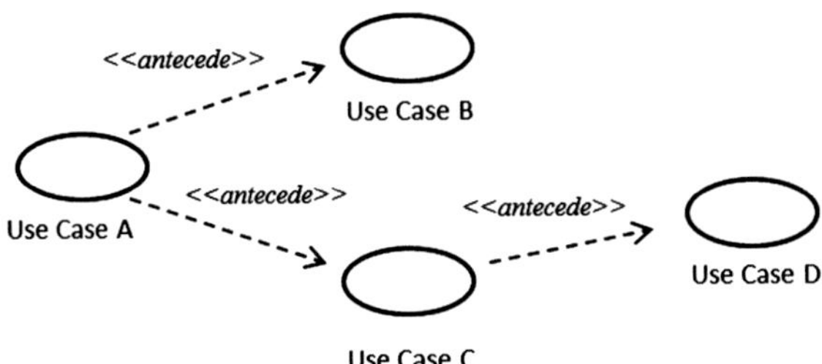

Fig. 1 Use case precedence diagram

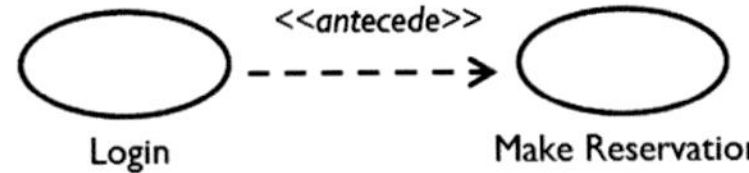

Fig. 2 Precedence Rule 1—diagram example

Fig. 3 Precedence Rule 2—diagram example

It is important to note that in the UCPD, Included and Extended Use Cases have not been considered since they can be part of other use cases that refer to them. Based on this UCPD, a construction sequence is defined. The use cases that are on the left side of the diagram will be implemented before the ones that are on the right side. For instance, in Fig. 1, "Use case A" will be implemented before "Use case C".

In [4, 5] it was included the results of controlled experiments in which UCPD is applied in case studies by practitioners, and undergraduate students. The obtained results show that UCPD has more significant advantages over the utilization of ad hoc techniques.

Although the results obtained in controlled experiments were satisfactory, there is a need also to assess users' response to the new procedure and their intention to use it in the future, for which reason it was applied at the end of the experiment a questionnaire based on the Method Adoption Model (MAM) [6].

The Method Evaluation Model (MEM) was proposed by Moody [6] and this model is an adaptation of the Technology Acceptance Model [7, 8] defined by Davis. MEM explains and predicts the adoption of methods (see Fig. 4).

The central constructs of the MEM constitute the Method Adoption Model (MAM) as it can be observed in Fig. 4. Those constructs are the following:

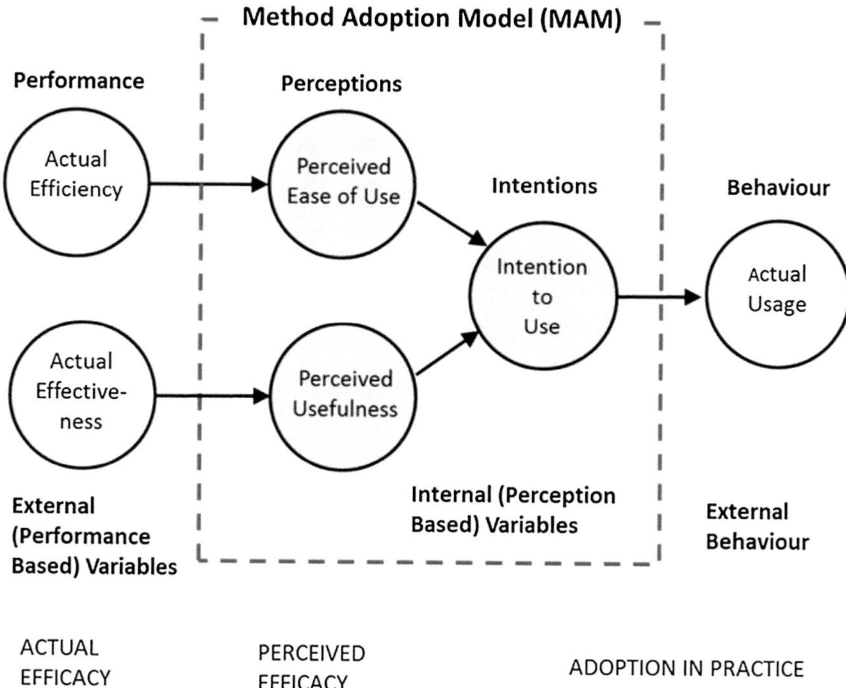

Fig. 4 The method evaluation model (MEM) [6]

- *Perceived Ease of Use (PEU)*: the extent to which a person believes that using a particular method would be effort-free.
- *Perceived Usefulness (PU)*: the extent to which a person believes that a particular method will be effective in achieving the intended objectives.
- *Intention to Use (IU)*: the extent to which a person intends to use a particular method.

MAM defines that perceived usefulness is influenced by the perceived ease of use, and intention to use is defined by perceived usefulness and perceived ease of use. Many empirical studies that evaluate software methods have been carried using MAM with students and practitioners [9–11]. Some of them do not confirm the relationships between the constructs defined by the MAM.

The rest of the paper is organized as follows: Sect. 2 describes this study, Sect. 3 details the results obtained for the empirical study. Finally, in Sect. 4, a summary and the plans for future research will conclude this paper.

2 Description of This Study

Using the Goal/Question/Metric (GQM) template for goal-oriented software measurement [12], it was defined this study as follows: **Analyze**: user's responses and effectiveness **For the purpose of**: evaluate **With respect to**: intention to use UCPD **In the context of**: undergraduate students and practitioners with at least 2 years of experience in software development projects, considering that the developer is free to select the sequence to construct use cases (there are no user's constraints).

Based on the MEM, it was formulated the working hypotheses of this research which make reference to the intrinsic constructs of the model. These hypotheses are stated in the following way:

- Hypothesis 1 (H_1): UCPD is perceived easy to use.
- Hypothesis 2 (H_2): UCPD is perceived useful.
- Hypothesis 3 (H_3): There is an intention to use UCPD in future software projects.
- Hypothesis 4 (H_4): The perceived ease of use (PEU) has a positive effect on the perceived usefulness (PU).
- Hypothesis 5 (H_5): The perceived ease of use (PEU) has a direct and positive effect on intention to use (IU).
- Hypothesis 6 (H_6): The perceived usefulness (PU) has a direct and positive effect on intention to use (IU).
- Hypothesis 7 (H_7): The actual effectiveness has a direct and a positive effect on perceived usefulness (PU).

2.1 Participants

The undergraduate students who participated in this study were fourth year students of the Informatics Engineering program at the Pontificia Universidad Católica del Perú (PUCP) that were enrolled in the Spring '06 and Fall '15.

The practitioners were 25 professionals with at least 2 years of experience who participated in a controlled experiment in 2007 and the quantitative results are detailed in [5]. This experiment was replicated in 2009 with 17 practitioners (with at least 2 years of experience too) who were graduate students of the Master in Informatics program at PUCP.

2.2 Materials

It was designed a questionnaire which included one question for each construct of the MAM. Each answer had to be quantified on a five point Likert-type scale [13]. It could be considered as a disadvantage to use only one question for each construct, but there are some studies that have applied this same approach in other fields such as the medicine with appropriate results [14–16]. The purpose was to create a user-friendly questionnaire.

The undergraduate students and the practitioners filled the questionnaire at the end of the controlled experiment in which they were involved into. It was commented to the participants that the purpose of the questionnaire is to know their honest opinion about UCPD.

Further details of the questionnaire used and the instruments utilized in the controlled experiment with practitioners can be found at: http://inform.pucp.edu.pe/~jpowsang/UCPD/mem.html.

2.3 Tasks Performed During the Experiment

The participants had to apply the first case study with their ad hoc techniques and the second one with UCPD. The participants had to applied two different case studies with similar characteristics (both are information systems) in order to mitigate the learning effects. Table 1 shows the tasks carried out in the session by the participants.

Unlike the experiment presented in [1], participants did not apply UCPD to the case study 1. For statistical purposes, it is only needed that students had to apply the technique to the case study 2.

The session lasted approximately 1 h and the practitioners performed 45 min on average to complete all the tasks. Even though it was not part of this study to know

Table 1 Tasks carried out by the subjects

Task Nº	Description
1	Receive case study 1 and questionnaire 1
2	Fill in questionnaire 1
3	Receive case study 2 and questionnaire 2
4	Elaborate use case precedence diagram for case study 2
5	Fill in questionnaire 2
6	Fill in questionnaire 3

which technique demanded less time, we could observe that they spent less than 10 min to elaborate UCPD.

3 Results

The statistical hypotheses to test the working hypothesis H_1, H_2, and H_3 are the following:

$$H_o : \mu \leq 3, \alpha = 0.05$$
$$H_a : \mu > 3$$

"μ" is the mean response obtained in the questions related to user's perception about UCPD. It can be considered a positive perception of the participants, if the mean response is greater than 3, because a five point Likert-type scale [13] was used in the questionnaries from 1 to 5. To evaluate the MEM relationships, correlation coeficients and regression analysis were used to formally test hypotheses H_4, H_5, H_6, and H_7.

3.1 Perceived Ease of Use (PEU)

Table 1 presents the results obtained with the question relate to perceived ease of use. It was established a significance level of 0.05 to statistically test the obtained results with undergraduate students and practitioners (Table 2).

In order to determine if the obtained results followed a normal distribution, it was applied the Shapiro-Wilk test [17]. Since the computed p-values were lower than the significance level $\alpha = 0.05$, the normal distribution hypothesis was rejected for the four samples (undergraduate students and practitioners). Due to these results, the Wilcoxon signed rank test [18] was chosen to test the statistical hypothesis defined previously (Ho: $\mu \leq 3$, Ha: $\mu > 3$). Since the computed p-values were lower than the significance level $\alpha = 0.05$, the null hypothesis Ho had to be rejected for all

Table 2 Descriptive statistics for perceived ease of use

Variable	Undergrad. students I	Undergrad. students II	Practitioners I	Practitioners II
Observations	35	19	25	14
Minimum	3	2	2	3
Maximum	5	5	5	5
Mean	3.971	4.105	3.84	4
Std. Dev.	0.618	0.936	0.850	0.679

of the samples. It means that it can be empirically corroborated the working hypothesis H_1: the undergraduate students and the practitioners perceived UCPD as easy to use.

3.2 *Perceived Usefulness (PU)*

Table 3 presents the results obtained with the question related to to usefulness. It was established a significance level of 0.05 to statistically test the obtained results. Similar to PEU, it was applied the Shapiro-Wilk test, and the normal distribution hypothesis was rejected for all of the samples. Using the Wilcoxon signed rank test in order to test the statistical hypothesis defined previously (Ho: $\mu \leq 3$, Ha: $\mu > 3$), the computed p-values were lower than the significance level $\alpha = 0.05$, and the null hypothesis Ho had to be rejected. It means that it can be empirically corroborated the working hypothesis H_2: the undergraduate students and the practitioners perceived UCPD as useful.

3.3 *Intention to Use (IU)*

Table 4 presents the results obtained with the question relate to intention to use. It was established a significance level of 0.05 to statistically test the obtained results.

Similar to the results obtained with PEU, and PU, samples did not follow a normal distribution either. Since the computed p-values using the Wilcoxon signed

Table 3 Descriptive statistics for perceived usefulness

Variable	Undergrad. students I	Undergrad. students II	Practitioners I	Practitioners II
Observations	35	19	25	14
Minimum	3	4	3	3
Maximum	5	5	5	5
Mean	4.486	4.473	4.36	4.214
Std. Dev.	0.658	0.512	0.638	0.699

Table 4 Descriptive statistics for perceived usefulness

Variable	Undergrad. students I	Undergrad. students II	Practitioners I	Practitioners II
Observations	35	19	25	14
Minimum	2	3	1	3
Maximum	5	5	5	5
Mean	4.4	4.315	4.2	4.214
Std. Dev.	0.736	0.749	0.913	0.579

rank test were lower than the significance level $\alpha = 0.05$, the null hypothesis Ho had to be rejected. It means that it can be can empirically corroborated the working hypothesis H_3: the undergraduate students and the practitioners have the intention to use UCPD.

3.4 MEM Evaluation

To assess the relationships between variables proposed in the MEM, it must be used the correlation coefficient, similar to the studies conducted by Davis [7] and Adams et al. [8].

According Muijs [19] in order to determine if there is a degree of relationship between two ordinal variables, the Spearman's correlation coefficient must be used (not the Pearson's correlation one). The Likert-scale used in the questionnaires is ordinal, for this reason Spearman's correlation (Spearman's r) had to be used to evaluate MAM. The rules of thumb to determine the strength of a relationship proposed by Muijs are the following:

<0. ±1 weak
<0. ±3 modest
<0. ±5 moderate
<0. ±8 strong
≥±0.8 very strong.

In the controlled experiments, it could be empirically corroborated that UCPD produced more accurate assessments than ad hoc techniques 16. In order to determine a possible relationship between effectiveness and usefulness, it was considered three levels of effectiveness comparing ad hoc vs. UCPD: (1) ineffectiveness, (2) equal effectiveness, (3) better effectiveness.

Table 5 presents Spearman's r (ρ) and the strength for each relationship for the undergraduate students' samples.

It can be observed that usefulness (PU) and ease of use (PEU) were significantly correlated each other for one sample. Usefulness (PU) and intention to use (IU) were significantly correlated each other for one sample. Ease of use (PEU) and intention to use (IU) were not only significantly correlated each other in both samples. Effectiveness and usefulness were significantly correlated each other in

Table 5 Strength for the MEM's relationship for undergraduate students

Characteristic	Undergrad. students I	Undergrad. students II
PEU versus PU	Modest	Moderate
	($\rho = 0.118$, p-value = 0.501)	($\rho = 0.461$, p-value = 0.0471)
PEU versus IU	Modest	Modest
	($\rho = 0.205$, p-value = 0.235)	($\rho = 0.104$, p-value = 0.672)
PU versus IU	Moderate	Moderate
	($\rho = 0.419$, p-value = 0.012)	($\rho = 0.336$, p-value = 0.160)
Effectiveness versus PU	Moderate	Very strong
	($\rho = 0.303$, p-value = 0.179)	($\rho = 1$, p-value = 0)

Table 6 Strength for the MEM's relationship for practitioners

Characteristic	Practitioners I	Practitioners II (master students)
PEU versus PU	Moderate	Strong
	($\rho = 0.416$, p-value = 0.039)	($\rho = 0.624$, p-value = 0.017)
PEU versus IU	Moderate	Moderate
	($\rho = 0.386$, p-value = 0.057)	($\rho = 0.368$, p-value = 0.195)
PU versus IU	Moderate	Strong
	($\rho = 0.318$, p-value = 0.122)	($\rho = 0.616$, p-value = 0.019)
Effectiveness versus PU	Moderate	Strong
	($\rho = 0.419$, p-value = 0.035)	($\rho = 741$, p-value = 0.002)

one sample. It means that it can be partially corroborated the working hypotheses H_4, H_6, and H_7 using both samples of undergraduate students. H_5 can not be confirmed with both samples.

Table 6 presents Spearman's r (ρ) and the strength for each relationship for samples of practitioners. It can be observed that usefulness (PU) and ease of use (PEU) were significantly correlated each other for both samples. Ease of use (PEU), and intention to use (IU) were significantly correlated each other for one sample. Usefulness (PU) and intention to use (IU) were also significantly correlated each other for one sample. Effectiveness and usefulness (PU) were significantly correlated each other for both samples. It means that it can be empirically corroborated the working hypotheses H_4, and H_7 using both samples of practitioners. H_5 and H_6 can be partially corroborated with the samples of practitioners.

In order to confirm the causal relationships between variables defined in the MAM, an ordinal regression analysis had to be used 20. All of the causal relationships could be confirmed using ordinal regression [20] (detailed information of these ordinal regression models are not included in this paper).

4 Conclusions and Future Work

This paper describes an empirical study that evaluates the intention to use the UCPD technique that is used to determine software construction sequences taking into account the developer's perspective. The study considers the perceptions of undergraduate students and practitioners with at least 2 years of experience in software development projects.

UCPD is perceived as easy to use and useful for all of the participants (undergraduate students and practitioners). Also, the participants of this study acknowledged having the intention to use UCPD in next software development projects.

Although the perceptions of UCPD are positive for all of the participants, the relationships defined in the MEM are best confirmed with the statistical tests applied to two samples of practitioners.

Many researchers comment the benefits to use undergraduate students for research studies [21]. However, it should be noted that in some situations, similar to this study, the results obtained with undergraduate students should be taken with caution and it is preferable to use practitioners with experience in the industry, in order to get confident results.

As a future work, it is planned to replicate the controlled experiment in order to contrast and confirm the results obtained in this study.

References

1. Pow-Sang, J.A., Nakasone, A., Imbert, R., Moreno, A.M.: An approach to determine software requirement construction sequences based on use cases. In: Proceedings Advanced Software Engineering and Its Applications-ASEA 2008 (Sanya, China), IEEE Computer Society (2008)
2. Jacobson, I.: Object-Oriented Software Engineering. A Use Case Driven Approach. Addison-Wesley, Reading (1992)
3. Rosenberg, D., Scott, K.: Use Case Driven Object Modeling with UML. Addison-Wesley, Massachusets (1999)
4. Pow-Sang, J.A., Imbert, R., Moreno, A. M.: A replicated experiment with undergraduate students to evaluate the applicability of a use case precedence diagram based approach in software projects. Commun. Comput. Inf. Sci. **257**, 169–179 (2011), Springer-Verlag
5. Pow-Sang, J.A., Técnicas para la Estimación y Planificación de Proyectos de Software con Ciclos de Vida Incremental y Paradigma Orientado a Objetos, PhD thesis, Universidad Politécnica de Madrid, 2012. http://oa.upm.es/10266/2/tesis-final-japowsang.pdf
6. Moody, D.L.: Dealing with complexity: a practical method for representing large entity relationship models, PhD. Thesis, Department of Information Systems, University of Melbourne, Australia (2001)
7. Davis, F.D.: Perceived usefulness, perceived ease of use and user acceptance of information technology. MIS Q. **13**(3), 319–340 (1989)
8. Adams, D., Nelson, R., Todd, P.: Perceived usefulness, ease of use, and usage of information technology: a replication. MIS Q. **16**(2), 227–247 (1993) (USA)
9. Abrahão, S.: On the functional size measurement of object-oriented conceptual schemas: design and evaluation issues, PhD Thesis, Department of Information Systems and Computation, Valencia University of Technology, October 2004

10. Condori, N.: Un Procedimiento de Medición de Tamaño Funcional para Especificaciones de Requisitos, PhD Thesis, Department of Information Systems and Computation, Valencia University of Technology (2007)
11. Poels, G., Maes, A., Gailly, F., Paemeleire, R.: Measuring user beliefs and attitudes towards conceptual schemas: tentative factor and structural equation model. In: Fourth Annual Workshop on HCI Research in MIS, December 2005
12. Basili, V.R., Caldiera, G., Rombach, H.D.: Goal question metric paradigm. In: Marciniak, J. J. (ed.) Encyclopedia of Software Engineering. Wiley, New York (1994)
13. Likert, R.: A Technique for the Measurement of Attitudes. Archives of Psychology. Columbia University Press, New York (1931)
14. Cepeda, M.S., Chapman, C.R., Miranda, N., Sanchez, R., Rodriguez, C.H., Restrepo, A.E., Ferrer, L.M., Linares, R.A., Carr, D.B.: Emotional disclosure through patient narrative may improve pain and well-being: results of a randomized controlled trial in patients with cancer pain. J. Pain Symptom Manage. **35**(6), 623–631 (June 2008), Elsevier
15. Davey, H.M., Barratt, A.L., Butow, P.N., Deeks, J.J.: A one-item question with a Likert or visual analog scale adequately measured current anxiety. J. Clin. Epidemiol. **60**, 356–360 (2007), Elsevier
16. Temel, J.S., Pirl, W.F., Recklitis, C.J.: Feasibility and validity of a one-item fatigue screen in a thoracic oncology clinic. J. Thoracyc Oncol. **1**(5) (June 2006), Lippincott Williams & Wilkins
17. Shapiro, S., Wilk, B.: An Analysis of variance test for normality (complete samples). Biometrika **52**(3/4), 591–611 (1965). http://www.jstor.org/stable/2333709
18. Wilcoxon, F.: Individual comparisons by ranking methods. Biometrics Bull. **1**(6), 80–83 (Dec 1945). http://www.jstor.org/stable/3001968
19. Muijs, D.: Doing Quantitative Research in Education with SPSS. Sage Publications, London (2004)
20. McCullagh, P., Nelder, J.A.: Generalized Linear Models, 2nd edn. Chapman & Hall, London (1989)
21. Carver, J., Jaccheri, L., Morasca, S.: Issues in using students in empirical studies in software engineering education. In: METRICS'03, p. 239, IEEE Computer Society, USA (2003)

Search-Based Software Engineering to Construct Binary Test-Suites

Jose Torres-Jimenez and Himer Avila-George

Abstract Search-based software engineering is the application of optimization techniques in solving software engineering problems. One challenge to testing software systems is the effort involved in creating test suites that will systematically test the system and reveal faults in an effective manner. Given the importance of the software testing phase, a specific subarea called search-based software testing has become increasingly important. This paper presents a search-based software testing tool (SBSTT), for constructing test suites. Through the use of SBSTT we were able to find 370 new upper bounds for binary test suites.

Keywords Search-based software testing · Combinatorial testing · Binary test-suites · Covering arrays

1 Introduction

In 2001, the term search-based software engineering (SBSE) was coined by Harman and Jones [1]. SBSE is the application of optimization techniques in solving software engineering problems. The applicability of optimization techniques in solving software engineering problems is suitable, as these problems frequently

J. Torres-Jimenez (✉)
Information Technology Laboratory, CINVESTAV-Tamaulipas, Ciudad Victoria, Tamaulipas, Mexico
e-mail: jtj@cinvestav.mx

H. Avila-George (✉)
Haramara TIC-LAB, CICESE-UT, Andador 10, Ciudad del Conocimiento, Tepic, Nayarit, Mexico
e-mail: himerag@cicese.mx; havilage@conacyt.mx

H. Avila-George
Cátedras CONACyT, Avenida Insurgentes Sur 1582, Benito Juárez, Crédito Constructor, 03940 Ciudad de, DF, Mexico

J. Mejia et al. (eds.), *Trends and Applications in Software Engineering*, Advances in Intelligent Systems and Computing 405,
DOI 10.1007/978-3-319-26285-7_17

require near optimal solutions. Search-based software testing (SBST) research has attracted much attention in recent years as part of a general interest in SBSE approaches [2]. The growing interest in SBST can be attributed to the fact that exhaustive testing is unfeasible considering the size and complexity for realistic or real-world software under test (SUT).

The main aim of software testing is to detect as many faults as possible in the SUT. To gain sufficient confidence that most faults are detected, exhaustive testing is an option. Since, in practice, this is not possible, testers resort to test models and coverage criteria to construct effective test cases that can reveal faults. Combinatorial interaction testing is a method that can reduce costs and increase the effectiveness of software testing [3]. It is based on constructing economical sized test-suites that provide coverage of the most prevalent combinations of factors.

A covering array is a combinatorial structure, denoted by `CA(N;t,k,v)` which can be described as a matrix with $N \times k$ elements, such that every $N \times t$ subarray contains all v^t possible combinations of symbols at least once. The parameter N is the number of rows, the parameter k is the number of factors (also called the degree), the parameter v is the alphabet (also called the order), and the parameter t is the strength of the coverage of interactions.

In order to clarify the use of covering arrays for software testing, consider the following example. Skype is one of the most popular Internet telephony service providers. The Fig. 1 shows the software component designed to manage the privacy settings provided by Skype. This software component has 12 parameters each one with 2 possible values (see Table 1). To test all the possible combinations of these parameters (i.e. exhaustive testing) we will need a set of $2^{12} = 4{,}096$ test cases. On the other hand, pairwise testing (i.e., using a covering array of strength 2) requires only 7 test cases.

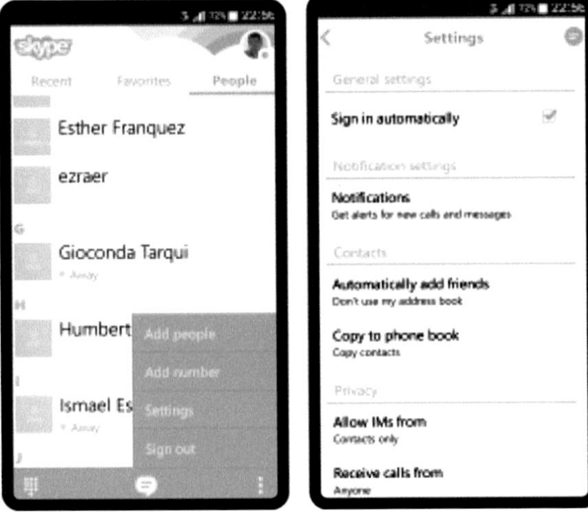

Fig. 1 Skype: privacy settings

Table 1 Skype configuration involves 12 parameters each one with two possible values

	Parameter	Value 1	Value 2
1	Sign in automatically	Checked	Unchecked
2	Notifications	Enable	Disable
3	Automatically add friends	Use my address book	Don't use my address book
4	Copy to phone book	Copy contacts	Don't copy contacts
5	Allow IMs from	Contacts only	Anyone
6	Receive calls from	Contacts only	Anyone
7	Allow Microsoft targeted ads	Checked	Unchecked
8	Answer call automatically	Checked	Unchecked
9	Enable video calling	Checked	Unchecked
10	Technical info	Checked	Unchecked
11	Video Quality	High	Low
12	Skype Wi-Fi	Enable	Disable

By labeling the values for each parameter as {0,1} the pairwise test suite can be represented by a covering array for 12 parameters with two values each one i.e. a CA(7;2,12,2). Any two columns, selected in any order, contain the order subsets of the symbols {0,1}, see Fig. 2.

In the CA(7;2,12,2), every row is a test case, the columns represent the parameters, and the symbol in each cell is the value of each parameter in a specific test case. This covering array ensures that if a failure is triggered with a particular combination of values in a pair of parameters, this combination is within the test suite. The interpretations of these input configurations are given in Table 2.

The *covering array number* is the minimum number of rows with which a specific covering array with parameters (t,k,v) exists, and is denoted by $CAN(t,k,v) = \min\{N : \exists CA(N;t,k,v)\}$. Given the values of t, k and v, the optimal covering array construction problem (CAC) consists in constructing a *CA(N;t,k,v)* such that the value of *N* is minimized. The determination of *CAN(t,k,v)* has been object of study and application in different research areas [4–7], but significantly the area with the most application of these objects is in software testing [8].

Covering arrays construction is a challenging combinatorial problem for which much research has been carried out in developing effective methods for constructing them. In this paper, we present a SBST tool (SBSTT) for constructing binary test suites.

SBSTT is further improvement based on previous works proposed by Avila-George et al. [9]. We use SBSTT to match or improve the size of binary

$$\begin{pmatrix} 0&1&1&1&1&1&0&0&0&0&0&1\\ 0&1&0&0&0&0&0&1&0&1&0&1\\ 1&1&0&1&0&0&1&0&1&1&1&0\\ 0&0&1&0&1&1&1&0&1&1&0&0\\ 1&0&1&0&1&0&0&1&1&0&0&1\\ 1&0&0&0&1&1&0&1&0&0&1&0\\ 0&0&1&1&0&1&1&1&0&0&1&1 \end{pmatrix}$$

Fig. 2 A covering array CA(7;2,12,2)

Table 2 Test suite based on Fig. 2

	Sign in automati cally	Notifications	Automations cally add friends	Copy to phone book	Allow IMs from	Receive calls from	Allow microsoft targeted	Answer call automati	Enable video calling	Technical info	Video quality	Skype Wi-Fi
1	Checked	Disable	Don't use my address book	Don't copy contacts	Anyone	Anyone	Checked	Checked	Checked	Checked	High	Disable
2	Checked	Disable	Use my address book	Copy contacts	Contacts only	Contacts only	Checked	Unchecked	Checked	Unchecked	High	Disable
3	Unchecked	Disable	Use my address book	Don't copy contacts	Contacts only	Contacts only	Unchecked	Checked	Unchecked	Unchecked	Low	Enable
4	Checked	Enable	Don't use my address	Copy contacts	Anyone	Anyone	Unchecked	Checked	Unchecked	Unchecked	High	Enable
5	Unchecked	Enable	Don't use my address	Copy contacts	Anyone	Contacts only	Checked	Unchecked	Unchecked	Checked	High	Disable
6	Unchecked	Enable	Use my address book	Copy contacts	Anyone	Anyone	Checked	Unchecked	Checked	Checked	Low	Enable
7	Checked	Enable	Don't use my address	Don't copy contacts	Contacts only	Anyone	Unchecked	Unchecked	Checked	Checked	Low	Disable

covering arrays reported by Colbourn [10] for $3 \leq t \leq 6$. Through the use of SBSTT we were able to find 370 new upper bounds. The constructed covering arrays are available at CINVESTAV Covering Array Repository (CCAR), which is available under request at http://www.tamps.cinvestav.mx/~jtj/. The final idea is to provide to the scientific community the best binary covering arrays without the need to waste any computer resource.

The remainder of this paper is organized in four more sections. In Sect. 2, a brief review is given to present the most representative metaheuristic techniques used to construct covering arrays. In Sect. 3, the SBSTT is explained in detail. In Sect. 4, the complete results for binary covering arrays will be shown. Final remarks are presented in the Sect. 5.

2 SBST—Test Suites Construction Methods

In recent years, many researchers have been interested in the use of metaheuristic search techniques for the construction of test suites. Search base software testing in the context of application of metaheuristics is related with construction of covering arrays, which are combinatorial objects. Given that in this work we deal only with binary covering arrays we recommend to the reader the review of the work proposed by Lawrence et al. [11], they presented a wide list of general algorithms to build binary covering arrays. In this section, we give a brief review of some metaheuristics for constructing covering arrays.

With respect to the application of metaheuristics, the objective function used for constructing a covering array is the number of uncovered t-tuples, so the covering array itself will have a cost of 0. Since one does not know the size of the test suite a priori, therefore, heuristic search techniques apply transformations to a fixed size array until all t-tuples are covered [12].

In 2001, Stardom [13] used simulated annealing (SA), genetic algorithms (GA) and tabu search (TS) for constructing covering arrays. The results indicated that SA and TS were best in constructing covering arrays. The GA implementation was the least effective; taking more time and moves to find good covering arrays. Stardom reported new upper bounds on size of covering arrays using SA. However, some of them were improved by Cohen et al. [14] in 2003. A simulated annealing implementation for constructing covering arrays was reported by Cohen et al. in [9]. The results showed that in comparison with greedy search algorithms used in test case generator (TCG) [15] and automatic efficient test generator (AETG) [16], SA improved on the bounds of minimum test cases in a test suite of strength two. In case of strength 3 constructions, the SA algorithm did not perform as well as the algebraic constructions. Therefore, the initial results indicated SA as more effective than other approaches for finding smaller sized test suites.

In 2004, Shiba et al. [17] used GA and ant colony algorithm (ACA) for constructing covering arrays. The results were compared with AETG [16], in-parameter order (IPO) [18] algorithm and SA. SA outperformed their results of using GA and

ACA with respect to the size of resulting test sets for two-way and three-way testing. However, their results outperformed AETG for two-way and three-way testing.

In 2004, Nurmela [19] presented a TS algorithm for building covering arrays. Nurmela's algorithm was able to improve some of the best-known upper bounds. However, a drawback of this algorithm is that it consumes considerably more computation time than any previously known greedy algorithm.

In 2007, Bryce and Colbourn [20] introduced a hybrid approach for constructing covering array. The hybrid approach applied a *one-test-at-a-time* greedy algorithm to initialize tests and then applied heuristic search to increase the number of t-tuples in a test. The heuristic search techniques applied were hill climbing, SA, TS and great flood. With different inputs, SA in general produced the quickest rate of coverage with 10 or 100 search iterations. Later in 2008, Cohen et al. [21] proposed a hybrid metaheuristic called Augmented Annealing. It employs recursive and direct combinatorial constructions to produce small building blocks, which are then augmented with a simulated annealing algorithm to construct a covering array. This method has been successfully used to construct covering arrays that are smaller than those created by using their simple SA algorithm.

In 2009, Walker II and Colbourn [22] proposed another TS. It utilizes a compact search space afforded by covering perfect hash families. Using this technique, improved covering arrays of strength 3, 4 and 5 have been found, as well as the first arrays of strength 6 and 7 found by computational search.

In 2010, Martinez-Pena el al. [23] proposed a SA algorithm for constructing ternary covering arrays. In this work, they represent the search space using trinomial coefficients.

In 2012, Torres-Jimenez and Rodriguez-Tello [24] introduced an improved implementation of a simulated annealing algorithm (ISA) for constructing binary covering arrays of different strengths. They have carried out extensive experimentation using a set of 127 benchmark instances taken from the literature. ISA algorithm was compared against the SA implementation published in [14, 21]. The results showed that ISA improves 11.11 % on average the best results found by others SA algorithms.

Recently, Avila-George et al. [25] presented a parallel SA for constructing ternary covering arrays. As a result of this, they presented 134 new upper bounds.

3 Search Based Software Testing Tool (SBSTT) Description

The SBSTT is an algorithm that works in two stages. In the first stage a smart initialization of a matrix of size $N \times k$ is performed, the idea is that the initial matrix containing the minimum number of missing t-tuples. In the second stage, the algorithm reduces to zero the missing t-tuples in a matrix.

ExtendCA is a greedy algorithm, which is responsible for making a smart initialization of an array of size $N \times k$, it receives as input a CA of size $N \times k - 1$. The strategy of the algorithm *ExtendCA* consists in filling the N cells of the new column one at time by trying all possible values in all free cells and choosing the pair cell/value that minimizes the number of missing combinations in the matrix.

The second stage is based on the enhanced simulated annealing (ESA) algorithm proposed by [9] but with the following differences: (1) SBSTT uses *ExtendCA* to construct an initial matrix, next it starts an incremental construction process. (2) The construction process is incremental in k, adding to a matrix as many columns as possible for the same value of N; and (3) SBSTT stops if a valid covering array is found or when 30 min have elapsed. The covering arrays construction process follows the next steps: (1) It must be defined the maximum number of rows (*MaxN*) and the maximum number of columns (*MaxK*); (2) It must specify the initial covering array; (3) Add a column to the input array covering (starts an instance of the algorithm *ExtendCA*); (4) Run an instance of the simulated annealing algorithm; (5) Check the number of missing combinations; (6) If the number of missing combinations is zero. Then, check if a new bound was achieved. If so, the new matrix is recorded. If it can be added one column, return to step 3. Otherwise, it finishes the algorithm; and (7) If the number of missing combinations is not zero. If it can be added a row to the matrix ($N < MaxN$), then the row is added to the matrix and it runs an instance of the simulated annealing algorithm, next back to step 5. Otherwise, it finishes the algorithm.

4 Results

This section presents an experimental design and results derived from the methodology described in the previous section. The purpose of SBSTT is to construct competitive test-suites. We propose the construction of binary covering arrays, for $3 \leq t \leq 6$. The detailed results produced by this experiment are listed in Tables 3, 4, 5 and 6. The results are grouped by t strength. β represents the new CAN and Δ shows the amount of new upper bounds.

Table 3 Results of strength $t = 3$

β	Δ
CAN(3,29,2) ≤ 23 ≤ CAN(3,30,2)	2
CAN(3,55,2) ≤ 29	1
CAN(3,57,2) ≤ 30	1
CAN(3,142,2) ≤ 39	1
CAN(3,145,2) ≤ 40 ≤ CAN(3,146,2)	2
CAN(3,157,2) ≤ 41 ≤ CAN(3,163,2)	8
CAN(3,178,2) ≤ 42 ≤ CAN(3,180,2)	3
CAN(3,258,2) ≤ 47 ≤ CAN(3,262,2)	5
CAN(3,263,2) ≤ 48 ≤ CAN(3,280,2)	18
Total	**41**

Table 4 Results of strength $t = 4$

β	Δ
CAN(4,34,2) ≤ 63	1
CAN(4,72,2) ≤ 78	1
CAN(4,79,2) ≤ 79	1
CAN(4,84,2) ≤ 87 ≤ CAN(4,89,2)	6
CAN(4,90,2) ≤ 92	1
CAN(4,91,2) ≤ 94 ≤ CAN(4,97,2)	7
CAN(4,98,2) ≤ 98 ≤ CAN(4,101,2)	4
CAN(4,102,2) ≤ 102 ≤ CAN(4,107,2)	6
CAN(4,108,2) ≤ 105 ≤ CAN(4,109,2)	2
CAN(4,110,2) ≤ 106 ≤ CAN(4,111,2)	2
CAN(4,112,2) ≤ 107 ≤ CAN(4,113,2)	2
CAN(4,114,2) ≤ 109	1
CAN(4,145,2) ≤ 119	1
CAN(4,146,2) ≤ 122	1
CAN(4,147,2) ≤ 125 ≤ CAN(4,149,2)	3
CAN(4,150,2) ≤ 126 ≤ CAN(4,151,2)	2
CAN(4,152,2) ≤ 127 ≤ CAN(4,155,2)	4
CAN(4,156,2) ≤ 128 ≤ CAN(4,157,2)	2
CAN(4,158,2) ≤ 129 ≤ CAN(4,160,2)	3
CAN(4,161,2) ≤ 130 ≤ CAN(4,165,2)	5
CAN(4,166,2) ≤ 131 ≤ CAN(4,169,2)	4
CAN(4,170,2) ≤ 134 ≤ CAN(4,172,2)	3
CAN(4,173,2) ≤ 135	1
CAN(4,174,2) ≤ 142 ≤ CAN(4,179,2)	6
CAN(4,180,2) ≤ 143 ≤ CAN(4,189,2)	10
CAN(4,190,2) ≤ 144 ≤ CAN(4,192,2)	3
CAN(4,193,2) ≤ 145 ≤ CAN(4,196,2)	4
CAN(4,197,2) ≤ 154 ≤ CAN(4,208,2)	12
CAN(4,209,2) ≤ 156 ≤ CAN(4,213,2)	5
CAN(4,214,2) ≤ 157 ≤ CAN(4,217,2)	4
CAN(4,218,2) ≤ 158 ≤ CAN(4,222,2)	5
CAN(4,257,2) ≤ 161 ≤ CAN(4,268,2)	12
CAN(4,273,2) ≤ 165 ≤ CAN(4,289,2)	17
CAN(4,290,2) ≤ 167 ≤ CAN(4,296,2)	7
CAN(4,297,2) ≤ 168 ≤ CAN(4,324,2)	28
Total	**176**

The analysis of the data presented led us to the following observation. The solutions quality attained by SBSTT is very competitive with respect to that produced by the state-of-the-art procedures reported in [10]. We were able to reach 370 new upper bounds.

Table 5 Results of strength $t = 5$

β	Δ
CAN(5,16,2) ≤ 98	1
CAN(5,17,2) ≤ 99 ≤ CAN(5,20,2)	4
CAN(5,21,2) ≤ 109	1
CAN(5,22,2) ≤ 111	1
CAN(5,23,2) ≤ 114	1
CAN(5,24,2) ≤ 117	1
CAN(5,25,2) ≤ 118	1
CAN(5,26,2) ≤ 130	1
CAN(5,73,2) ≤ 159 ≤ CAN(5,80,2)	8
CAN(5,81,2) ≤ 168 ≤ CAN(5,84,2)	4
CAN(5,85,2) ≤ 178 ≤ CAN(5,90,2)	6
CAN(5,91,2) ≤ 194 ≤ CAN(5,98,2)	8
CAN(5,99,2) ≤ 202 ≤ CAN(5,102,2)	4
CAN(5,103,2) ≤ 206	1
CAN(5,104,2) ≤ 208	1
CAN(5,105,2) ≤ 214 ≤ CAN(5,108,2)	4
CAN(5,109,2) ≤ 218 ≤ CAN(5,110,2)	2
CAN(5,111,2) ≤ 225 ≤ CAN(5,114,2)	4
CAN(5,115,2) ≤ 241 ≤ CAN(5,122,2)	8
CAN(5,123,2) ≤ 252 ≤ CAN(5,128,2)	6
CAN(5,129,2) ≤ 260 ≤ CAN(5,132,2)	4
CAN(5,133,2) ≤ 271 ≤ CAN(5,138,2)	6
CAN(5,139,2) ≤ 273 ≤ CAN(5,140,2)	2
CAN(5,141,2) ≤ 288 ≤ CAN(5,143,2)	3
CAN(5,144,2) ≤ 289 ≤ CAN(5,150,2)	7
CAN(5,151,2) ≤ 292 ≤ CAN(5,152,2)	2
CAN(5,153,2) ≤ 303 ≤ CAN(5,155,2)	3
CAN(5,156,2) ≤ 304 ≤ CAN(5,158,2)	3
CAN(5,159,2) ≤ 310 ≤ CAN(5,164,2)	6
CAN(5,165,2) ≤ 318 ≤ CAN(5,168,2)	4
CAN(5,169,2) ≤ 322 ≤ CAN(5,170,2)	2
CAN(5,171,2) ≤ 327 ≤ CAN(5,174,2)	4
CAN(5,175,2) ≤ 332 ≤ CAN(5,180,2)	6
CAN(5,181,2) ≤ 346 ≤ CAN(5,192,2)	12
CAN(5,193,2) ≤ 357 ≤ CAN(5,194,2)	2
Total	**133**

Table 6 Results of strength $t = 6$

β	Δ
CAN(6,18,2) ≤ 260	1
CAN(6,19,2) ≤ 285	1
CAN(6,20,2) ≤ 300	1
CAN(6,21,2) ≤ 318	1
CAN(6,22,2) ≤ 330	1
CAN(6,23,2) ≤ 350	1
CAN(6,24,2) ≤ 357	1
CAN(6,25,2) ≤ 375	1
CAN(6,26,2) ≤ 383	1
CAN(6,27,2) ≤ 404	1
CAN(6,28,2) ≤ 415	1
CAN(6,29,2) ≤ 426	1
CAN(6,30,2) ≤ 441	1
CAN(6,31,2) ≤ 457	1
CAN(6,32,2) ≤ 468	1
CAN(6,33,2) ≤ 480	1
CAN(6,34,2) ≤ 482	1
CAN(6,35,2) ≤ 496	1
CAN(6,36,2) ≤ 505	1
CAN(6,37,2) ≤ 514	1
Total	**20**

5 Conclusions

SBSE is the application of optimization techniques in solving software engineering problems. One challenge to testing software systems is the effort involved in creating test suites that will systematically test the system and reveal faults in an effective manner. Given the importance of the software testing phase, SBST has become increasingly important. The growing interest in SBST can be attributed to the fact that exhaustive testing is unfeasible considering the size and complexity for realistic or real-world software under test (SUT). Therefore, the testers have resorted to test models and coverage criteria to construct effective test cases that can reveal faults. Combinatorial interaction testing is a method that can reduce costs and increase the effectiveness of software testing. It is based on constructing test-suites of economical size, which provide coverage of the most prevalent configurations of parameters. A covering array is a combinatorial object, which can be used to represent these test-suites.

Binary covering arrays construction is a challenging combinatorial problem for which much research has been carried out in developing effective methods for constructing them. In this paper, a further advancement based on the algorithm proposed by Avila-George et al [9]. was presented. The purpose of this work was to construct small binary test-suites for $3 \leq t \leq 6$.

The quality solution attained by SBSTT is very competitive with respect to that produced by the state-of-the-art procedures. In fact, SBSTT algorithm was able to reach 370 new upper bounds.

We would like to point out that the covering arrays constructed using the methodology reported in this paper are available under request at the following address: http://www.tamps.cinvestav.mx/~jtj.

Acknowledgments The authors acknowledge GENERAL COORDINATION OF INFORMATION AND COMMUNICATIONS TECHNOLOGIES (CGSTIC) at CINVESTAV for providing HPC resources on the Hybrid Cluster Supercomputer "Xiuhcoatl", that have contributed to the research results reported. The following projects have funded the research reported in this paper: 51623 - Fondo Mixto CONACyT y Gobierno del Estado de Tamaulipas; 238469 - CONACyT Métodos Exactos para Construir Covering Arrays Óptimos; 232987 - CONACyT Conjuntos de Prueba Óptimos para Métodos Combinatorios (Optimal Test Sets for Combinatorial Methods); 2143 - Cátedras CONACyT.

References

1. Harman, M., Jones, B.F.: Search-based software engineering. Inf. Softw. Technol. **43**(14), 833–839 (2001)
2. Ali, S., Briand, L.C., Hemmati, H., Panesar-Walawege, R.K.: A systematic review of the application and empirical investigation of search-based test case generation. IEEE Trans. Softw. Eng. **36**(6), 742–762 (2010)
3. Kuhn, D.R., Lei, Y., Kacker, R.N.: Practical combinatorial testing: beyond pairwise. IT Prof. **10**(3), 19–23 (2008)
4. Cawse, J.N.: Experimental Design for Combinatorial and High Throughput Materials Development. Wiley, New York (2003)
5. Hedayat, A.S., Sloane, N.J.A., Stufken, J.: Orthogonal Arrays: Theory and Applications. Springer Science & Business Media, Berlin (1999)
6. Shasha, D.E., Kouranov, A.Y., Lejay, L.V., Chou, M.F., Coruzzi, G.M.: Using combinatorial design to study regulation by multiple input signals: a tool for parsimony in the post-genomics era. Plant Physiol. **127**(4), 1590–1594 (2001)
7. Vadde, K.K., Syrotiuk, V.R.: Factor interaction on service delivery in mobile ad hoc networks. IEEE J. Sel. Areas Commun. **22**(7), 1335–1346 (2004)
8. Avila-George, H., Torres-Jimenez, J., Gonzalez-Hernandez, L., Hernández, V.: Metaheuristic approach for constructing functional test-suites. IET Softw. **7**(2), 104–117 (2013)
9. Avila-George, H., Torres-Jimenez, J., Hernández, V.: Constructing real test-suites using an enhanced simulated annealing. In: Pavón, J., Duque-Méndez, N.D., Fuentes-Fernández, R. (eds.) Advances in Artificial Intelligence – IBERAMIA 2012, pp. 611–620. Springer, Berlin (2012)
10. Colbourn, C.J.: Covering array tables for t = 2,3,4,5,6. http://www.public.asu.edu/~ccolbou/src/tabby/catable.html. Accessed 1 July 2015
11. Lawrence, J., Kacker, R.N., Lei, Y., Kuhn, D.R., Forbes, M.: A survey of binary covering arrays. Electron J. Comb. **18**(1), 1–30 (2011)
12. Afzal, W., Torkar, R., Feldt, R.: A systematic review of search-based testing for non-functional system properties. Inf. Softw. Technol. **51**(6), 957–976 (2009)
13. Stardom, J.: Metaheuristics and the Search for Covering and Packing Arrays. Simon Fraser University, Burnaby (2001)

14. Cohen, M.B., Gibbons, P.B., Mugridge, W.B., Colbourn, C.J.: Constructing test suites for interaction testing. In: Proceedings of the 25th International Conference on Software Engineering, 2003, pp. 38–48 (2003)
15. Tung, Y.-W., Aldiwan, W.S.: Automating test case generation for the new generation mission software system. In: 2000 IEEE Aerospace Conference Proceedings, vol. 1, pp. 431–437 (2000)
16. Cohen, D.M., Dalal, S.R., Fredman, M.L., Patton, G.C.: The AETG system: an approach to testing based on combinatorial design. IEEE Trans. Softw. Eng. **23**(7), 437–444 (1997)
17. Shiba, T., Tsuchiya, T., Kikuno, T.: Using artificial life techniques to generate test cases for combinatorial testing. In: Proceedings of the 28th Annual International Computer Software and Applications Conference, pp. 72–77 (2004)
18. Lei, Y., Tai, K.-C.: In-parameter-order: a test generation strategy for pairwise testing. In: Proceedings of the Third IEEE International High-Assurance Systems Engineering Symposium, 1998, pp. 254–261 (1998)
19. Nurmela, K.J.: Upper bounds for covering arrays by tabu search. Discret. Appl. Math. **138**(1–2), 143–152 (2004)
20. Bryce, R.C., Colbourn, C.J.: The density algorithm for pairwise interaction testing. Softw. Test. Verif. Reliab. **17**(3), 159–182 (2007)
21. Cohen, M.B., Colbourn, C.J., Ling, A.C.H.: Constructing strength three covering arrays with augmented annealing. Discret. Math. **308**(13), 2709–2722 (2008)
22. Walker II, R.A., Colbourn, C.J.: Tabu search for covering arrays using permutation vectors. J. Stat. Planning Infer. **139**(1), 69–80 (2009)
23. Martinez-Pena, J., Torres-Jimenez, J., Rangel-Valdez, N., Avila-George, H.: A heuristic approach for constructing ternary covering arrays using trinomial coefficients. In: Kuri-Morales, A., Simari, G. (eds.) Advances in Artificial Intelligence—IBERAMIA 2010, vol. 6433, pp. 572–581. Springer, Berlin (2010)
24. Torres-Jimenez, J., Rodriguez-Tello, E.: New bounds for binary covering arrays using simulated annealing. Inf. Sci. **185**(1), 137–152 (2012)
25. Avila-George, H., Torres-Jimenez, J., Hernández, V.: New bounds for ternary covering arrays using a parallel simulated annealing. Math. Probl. Eng. **2012**(Article ID 897027), 19 (2012)

The Use of Simulation Software for the Improving the Supply Chain: The Case of Automotive Sector

Cuauhtémoc Sánchez Ramírez, Giner Alor Hernández, Jorge Luis García Alcaraz and Diego Alfredo Tlapa Mendoza

Abstract The effective decision making in the supply chain automotive can be complex because the number of suppliers and customers who form part of it. For this reason, is necessary the use of tools to help the stakeholder in the selection of better decisions that allows for improving the performance of the supply chain, one of these tools is the simulation. In this paper a system dynamics approach is used to develop a simulation model to evaluate three different scenarios that improve the performance of supply chain automotive.

Keywords Simulation software · Supply chain automotive · Decision making

1 Introduction

In a globalized market the automotive sector the assemblers and suppliers have been forced to improve their processes and products as well as to define strategies to enlarge their warranty time and to offer better after-sale services, so the strategies

C.S. Ramírez (✉) · G.A. Hernández
Division of Research and Postgraduate Studies, Instituto Tecnológico de Orizaba, Av. Oriente 9, 852. Col Emiliano Zapata, C.P. 94320 Orizaba, Veracruz, Mexico
e-mail: csanchez@itorizaba.edu.mx

G.A. Hernández
e-mail: galor@itorizaba.edu.mx

J.L.G. Alcaraz
Department of Industrial Engineering, Autónoma de Ciudad Juárez, Avenida del Charro, # 450 Norte Colonia Partido Romero, C.P. 32310 Ciudad Juárez, Chihuahua, Mexico
e-mail: jorge.garcia@uacj.mx

D.A.T. Mendoza
Engineering, Architecture and Design Faculty, Universidad Autónoma de Baja California, Carretera Transpeninsular Ensenada-Tijuana # 3917, Colonia Playitas, C.P. 22860 Ensenada, Baja California, Mexico
e-mail: diegotlapa@uabc.edu.mx

J. Mejia et al. (eds.), *Trends and Applications in Software Engineering*, Advances in Intelligent Systems and Computing 405,
DOI 10.1007/978-3-319-26285-7_18

based only on costs are no longer the base for a competitive advantage. In the automotive sector the supply chain management is a key factor to success due to the quantity of components needed to make a vehicle. According to [1] a supply chain as independent organizations network working in coordination to control, manage and improve the material flow and information from the raw material suppliers to the final customer.

The design and structure of supply chain is complex which makes necessary the use of different tools, methodologies and systems to improving the supply chain. One of these tools is the simulation. The Simulation is the process of exercising a model for a particular instantiation of the system and specific set of inputs in order to predict the system response [2].

In the automotive supply chain is submitted to many decisions, nevertheless, the variable number and key indicators to consider for the evaluation and development is much extended and complex. In decision making, the people only consider a part of them or the ones they believe are most important, causing a tendency to problem solution or limiting the best solution area, reducing that way the effectiveness of the decision [3]. One of the principal Key Performance Indicator (KPI) in the development of the supply chain is the *order fulfillment* and some variables that impact the performance of this indicator are the cycle time, inventory management and others.

This article presents a simulation model under approach in System Dynamics through a sensibility analysis evaluates the impact the following variables in the performance of supply chain automotive: *Cycle Time (T), Production Adjustment Time (PAT), Delivery Time (DT), Inventory Raw Material (IRM) and Inventory Finished Good (IFG)*. The case of study is a company member of the Automotive Cluster of Coahuila.

This chapter is structured by a *background section, development of the model section, and conclusions.*

2 Background

2.1 Supply Chain

The main objective of a supply chain management is to create initiatives to improve its development, diminishing costs and improving customer service, to accomplish this, it is necessary that suppliers, producers and distributors are in constant synchronization of all its activities. However, in the practice, this objective is not always possible to achieve, due to failure in the time agreed and scheduled, the variability of the cycle times in every process of the supply chain as well as the uncertainty that can affect the demand, which are factors that increase the complexity of the problem.

There are many techniques and tools used for analysing, evaluating and decision making in the chain supply management, one of these is the simulation. Simulation is an effective tool in production operation and logistic systems in which the main strength compared with math programming methods or stochastic models is that lets the user to observe, analyse and learn the dynamic behaviours system [4]. There are different types of simulation to evaluate the supply chain, five are the most important: (i) Simulation on Calculation Sheet; (ii) System Dynamics; (iii) Simulation Discrete Events; (iv) Dynamics Systems; and (v) Business Simulators.

The System Dynamics methodology has been selected in this article because is easy to represent delays that allow seeing the system behaviour under a study [5].

2.2 *Supply Chain with System Dynamics Approach*

The System Dynamics methodology was developed by Jay Forrester in the beginning of the 60s, because of its interdisciplinary focus the system dynamics helps to understand the dynamic characteristics of a complex system. The causal loop diagrams is the main tool of this methodology [6].

Some authors have developed simulation models using this methodology, where they consider as an important part the cycle time that affect de performance of supply chain and automotive sector. For example [7] developed a model to analyse planning strategies of resources on a supply chain; [8] propose strategies for a supply chain on the food industry where the transportation and delivery time is an important factor; [9] analysed the increasing capacity of a supply chain in which the life time of the product is short, considering among relevant variables the time adjustments of production for fast response to the demand due to life characteristics of the product. [10] Evaluated the impact of the Tohoku Earthquake as disruptions to the industrial supply chains in Japan at 2011. The flexibility in a multi-tier in a supply chain automotive is evaluated in [11, 12] Proposed a model to evaluate the inventory trends in emerging market. The reduction cost in the procurement processes automotive supply chain is analysed in [13].

3 Development of the Model

The automotive industry in Mexico is important for the national economy, because represents the 3.5 % of Gross Domestic Product (GDP) and 20 % employment of manufacturing [14]. Actually, Mexico is considered as an important member of the international network of automotive production. Important assemblers as Chrysler, Ford, General Motors, Nissan, Volkswagen and others have success operations in Mexico.

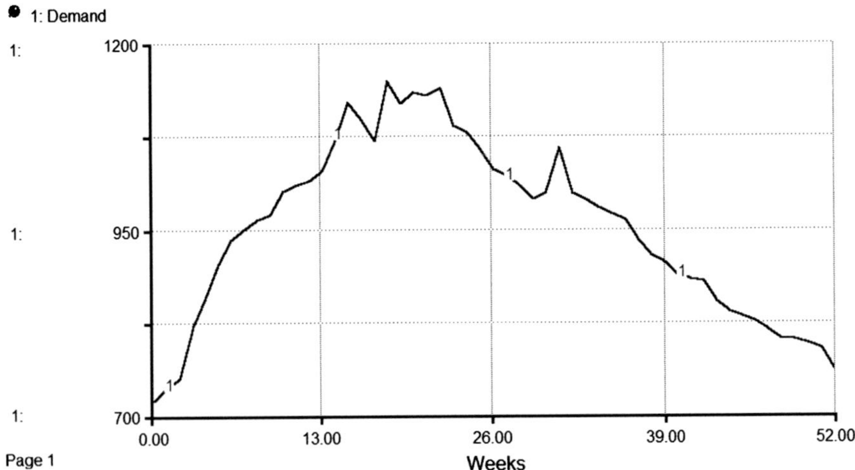

Fig. 1 Behaviour of demand on the next 52 weeks

The composition of automotive cluster from Coahuila is structured by two assemblers mainly and their Tier 1 suppliers, in which most of them are global corporations [15]. The case study was focus in the company called ABC that belongs to a supply chain of Automotive Cluster from Coahuila. Because the increase of customer demands, this company wants to analyse the affectation on this increase demand in the performance of its supply chain. Figure 1, shows the behaviour expected on the demand during the next 52 weeks.

3.1 Causal Loop Diagram

To structure the causal diagram the following authors were considered [8, 9, 16]. The model has been divided in three processes for a better comprehension: (1) Sourcing, (2) Production and, (3) Distribution and Evaluation. The processes are described below.
Sourcing:

- *Inventory of Raw Material*: Inventory level of available material for the production process.
- *Ordering to Supplier*: Indicates when a raw material order has been generated to the supplier.
- *Delivery Time*: Time that the supplier takes to put the order.

Production:

- *Production Capacity*: Maximum capacity that the company can produce with available resources.

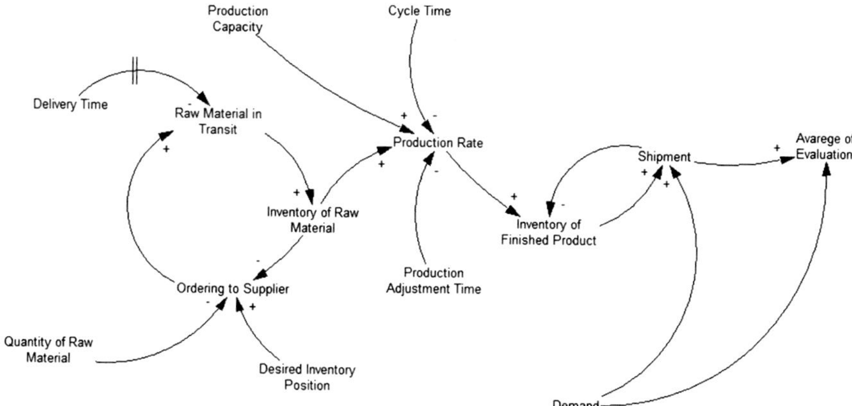

Fig. 2 Causal loop diagram

- *Cycle Time*: Time processing of the raw material in Finished Good.

Distribution and Evaluation:

- *Demand*: Customer requirements
- *Evaluation*: Difference between the requirements of the customers and what the company supplies of those requirements.
- *Inventory of Finished Good*: Inventory level of finished product used to supply the orders.
- *Shipment*: Orders supplied to customers.

In the Fig. 2 shows the main variables of the model through causal diagram. To understand the causal diagram is necessary to analyse the impact between variables. For example delivery time has an impact negative about the raw material in transit that means if delivery time increasing the raw material in transit will be reduced. On the other hand, if Raw Material in Transit decreasing the Inventory of Raw Material decreasing too o vice versa, due to positive relation represented by the positive sign on the arrow.

4 Results and Analysis

The model was developed in STELLA Version 10® and simulated during 52 weeks with the demand presented in Fig. 1. Among the assumptions considered in this model were the following, the order point and the quantity of raw material is constant when an order is generated; the raw material warehouse has a maximum capacity of 8500 units; the production capacity has a normal distribution behaviour with a mean of 1300 units and a standard deviation of 30 units; the warehouse of finished good has a capacity for 5000 units.

The Key Performance Indicator (KPI) to consider in the simulation model is the *fulfilment orders,* because measure the responsiveness of the company in compliance with customer orders.

Taking into account the data showing in Table 1, only have in average 79.5 % in the fulfillment orders, because Delivery Time of Raw Material are three weeks and has a great impact in performance of deliveries of company.

4.1 Analysis of Sensitivity and Results

As shown in Table 1, the company would not fulfillment the customer's orders if demand is the showing in Fig. 1. For this reason, in this section a sensibility analysis is presented with the objective to find the values in the processes which enable the total fulfillment orders. As discussed in the introduction five variables are considered to evaluate the performance of supply chain of the company ABC. Then, in this section the analysis of sensitivity is presented to improve the performance of the supply chain of the company. Two scenarios are proposed.

First scenario considers the following data:

Cycle Time (CT) = 1.5 weeks
Production Adjustment Time (PAT) = 1, 2 and 3 weeks
Delivery Time (DT) = 1, 2 y 3 weeks
Inventory Raw Material (IRM) = 1000, 2000 and 3000 units
Inventory Finished Good (IFG) = 1000, 2000 and 3000 units

Table 2 shows that the average fulfillment orders would increase to ***99.13 %*** with the following values *CT = 1.5, PAT = 1, DT = 1, IFG = 3000* and *IRM = 2000,* nevertheless, it can't be fulfillment the 100 % of the orders. The Cycle Time remains constant during the evaluation of the table because is a critical variable.

Table 1 Values and initial results of the model

Variables	Initial values	Results
Delivery time of raw material	3 weeks	*
Cycle time	1.5 weeks	*
Adjustment production time	3 weeks	*
Raw material warehouse	1500 units	6682 Units
Supplier lot	7500 units	*
Order point	2500 units	*
Finished good warehouse	1200 units	1001 Units
Number of ordering to supplier	0	6
Average fulfillment orders	0	**79.51 %**

***They remain constant**

Table 2 Results obtained with a cycle time 1.5 weeks

			CT = 1.5								
			PAT = 1			PAT = 2			PAT = 3		
			IRM			IRM			IRM		
			1000	2000	3000	1000	2000	3000	1000	2000	3000
DT = 1	IFG	1000	92.39	94.34	94.21	92.03	92.35	92.21	91.52	92.34	92.14
		2000	94.97	96.92	96.8	94.61	94.94	94.79	94.40	94.92	94.72
		3000	97.28	*99.13*	99.00	96.91	97.15	97.01	96.70	97.14	96.94
DT = 2	IFG	1000	85.55	87.73	88.34	85.50	87.68	88.27	84.21	86.39	88.21
		2000	88.13	90.31	90.93	88.08	90.26	90.86	86.79	88.97	90.79
		3000	90.43	92.54	93.13	90.38	92.5	93.07	89.09	91.21	93.01
DT = 3	IFG	1000	77.59	79.76	80.D0	77.54	79.71	79.97	77.48	79.66	79.92
		2000	80.17	82.35	82.58	80.12	82.3	82.55	80.07	82.25	82.51
		3000	82.47	84.58	84.79	82.42	84.53	84.76	82.37	84.48	84.72

Table 3 Results obtained with a cycle time 1 week

			CT = 1								
			PAT = 1			PAT = 2			PAT = 3		
			IRM			IRM			IRM		
			1000	2000	3000	1000	2000	3000	1000	2000	3000
DT = 1	IFG	1000	94.78	96.78	96.75	94.62	94.98	94.85	94.32	94.90	94.69
		2000	97.24	99.15	99.12	97.09	97.35	97.22	96.79	97.27	97.06
		3000	99.18	*100*	*100*	99.13	99.17	99.04	98.86	99.17	98.96
DT = 2	IFG	1000	37.33	89.61	90.22	87.29	89.58	90.16	85.91	88.2	90.11
		2000	89.79	92.02	92.59	89.76	91.99	92.53	88.38	90.61	92.48
		3000	99.02	93.94	94.41	91.99	93.93	94.4	90.61	92.59	94.4
DT = 3	IFG	1000	79.59	81.87	82.24	79.54	81.83	82.21	79.49	81.87	82.16
		2000	32.05	84.28	84.61	82.01	84.23	34.58	81.19	84.19	34.53
		3000	84.28	86.39	85.63	84.23	86.35	86.62	84.19	86.3	86.58

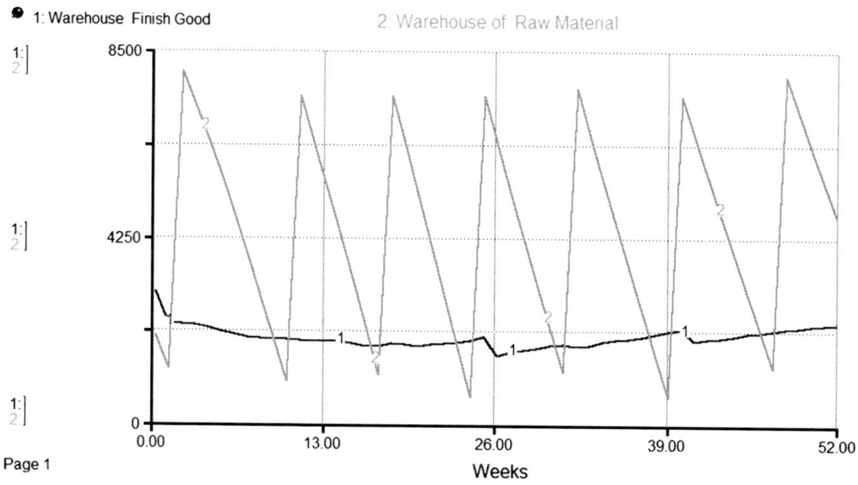

Fig. 3 Behaviour of warehouse of raw material and finished good

Second scenario considers the following data:

Cycle Time (CT) = 1 week
Production Adjustment Time (PAT) = 1, 2 and 3 weeks
Delivery Time (DT) = 1, 2 y 3 weeks
Inventory Raw Material (IRM) = 1000, 2000 and 3000 units
Inventory Finished Good (IFG) = 1000, 2000 and 3000 units

With the data proposed in the second scenario is possible to fulfillment the **100 %** of the orders (Table 3), with the following values *CT = 1, PAT = 1,DT = 1, IFG = 3000* and *IRM = 2000* and *CT = 1, PAT = 1,DT = 1,IFG = 3000* and *IRM = 3000.* Furthermore in Fig. 3 the Finished Good Inventory not presented disruption.

5 Conclusions and Future Work

The importance of simulation software is demonstrated in this case of study, due to the complexity of the supply chain automotive. One the most important KPI (Key Performance Indicator) in the supply chain automotive is the fulfilment orders, in this article is presented a simulation model to analyse the impact by increasing demand in the development in a company of automotive Cluster of Coahuila and the impact in this KPI. Sensitivity analysis is used to evaluate five critical variables and define best politics. To fulfilment of the total orders the following values must be met: *Cycle Time = 1 week, Production Adjustment Time = 1 week, Delivery Time = 1 week Inventory of Finished Good = 3000 units, Inventory of Raw*

Material = 2000 units. As future work it is proposed the analysis of others variables that improved the development the supply chain automotive. Also to include indicator cost.

Acknowledgements This work was supported by the National Council for Science and Technology of Mexico (CONACYT), National Technology of Mexico (TNM), and the Ministry of Public Education in Mexico (SEP) through PRODEP.

References

1. Christopher, Martin, Towill, Denis: An integrated model for the design of agile supply chains. Int. J. Phys. Distrib. Logistics Manage. **30**(4), 235–246 (2001)
2. Agte J., Borer N., de Weck O.: A simulation-based design model for analysis and optimization of multi-state aircraft performance. In: 6th AIAA Multidisciplinary Design Optimization Specialist Conference, pp. 12–15. Orlando, Florida (2010)
3. Oliva, R., Watson, N: What drives supply chain behavior? Research & Ideas. Harvard Business Review. http://hbswk.hbs.edu/item/4170.html (2007)
4. Umeda, S., Lee T.: Design specifications of a generic Supply Chain Simulator. In: Proceedings of the 2004 Winter Simulation Conference, pp. 1158–1166 (2004)
5. Angerhofer, B., Angelides, M.: System dynamics modelling in supply chain management: research view. In: Proccedings of the 2000 Winter Simulation Conference, pp. 342–351 (2000)
6. Sterman, J.: Business dynamics—systems thinking and modeling for a complex world. Mc Graw Hill, Boston (2000)
7. Ritchie-Dunham J. et al.: A strategic supply chain simulation model. In: Proccedings of the 2000 Winter Simulation Conference, pp. 1260–1264
8. Georgiadis, P., Vlachos, D., Iakovou, E.: A system dynamics modeling framework for a strategic suppl chain management of food chains. J. Food Eng. **70**, 351–364 (2004)
9. Kamath, N., Roy, R.: Capacity augmentation of a supply chain for a short lifecycle product: a system dynamics framework. Eur. J. Oper. Res. **179**, 334–351 (2007)
10. Matsuo, H.: Implications of the Tohoku earthquake for Toyota's coordination mechanism: supply chain disruption of automotive semiconductors. Int. J. Prod. Econ. **161**, 217–227 (2015)
11. Márcio, A., Scavarda, L.F., Pires, S., Ceryno, P., Klingebiel, K.: A multi-tier study on supply chain flexibility in the automotive industry. Int. J. Prod. Econ. **158**, 91–105 (2014)
12. Saranga, H., Mukherji, A., Shah, J.: Inventory trends in emerging market supply chains: evidence from the Indian automotive industry. IIMB Manage. Rev. **27**, 6–18 (2015)
13. Medeiros, G., Sellitto, M., Borchardt, M., Geiger, Albert: Procurement cost reduction for customized non-critical items in an automotive supply chain: an action research project. Ind. Mark. Manage. **40**, 28–35 (2011)
14. AMIA: www.amia.com.mx (2015)
15. Sánchez, C., Cedillo, M., Pérez, P., Martínez, J.: Global economic crisis and Mexican automotive suppliers: impacts on the labor capital. Simul. Trans. Soc. Model. Simul. Int. **87**, 711–725 (2011)
16. Forrester, J.: Industrial dynamics. Editorial Pegasus (1961)

App Reviews: Breaking the User and Developer Language Barrier

Leonard Hoon, Miguel Angel Rodriguez-García, Rajesh Vasa, Rafael Valencia-García and Jean-Guy Schneider

Abstract Apple, Google and third party developers offer apps across over twenty categories for various smart mobile devices. Offered exclusively through the App Store and Google Play, each app allows users to review the app and their experience with it. Current literature offers a general statistical picture of these reviews, and a broad overview of the nature of discontent of apps. However, we do not yet have a good framework to classify user reviews against known software quality attributes like performance or usability. In order to close this gap, in this paper, we develop an ontology encompassing software attributes derived from software quality models. This decomposes into approximately five thousand words that users employ to review apps. By identifying a consistent set of vocabulary that users communicate with, we can sanitise large datasets to extract stakeholder actionable information from reviews. The findings offered in this paper assists future app review analysis by bridging end-user communication and software engineering vocabulary.

Keywords Ontology development · Semantic annotation · Sentiment analysis · User reviews

L. Hoon · R. Vasa · J.-G. Schneider
Faculty of Science, Engineering and Technology, Swinburne University of Technology, Melbourne, Australia
e-mail: lhoon@swin.edu.au

R. Vasa
e-mail: rvasa@swin.edu.au

J.-G. Schneider
e-mail: jschneider@swin.edu.au

M.A. Rodriguez-García
Computational Bioscience Research Center, King Abdullah University of Science and Technology, 4700 KAUST, 1608, 23955-6900 Thuwal, Kingdom of Saudi Arabia
e-mail: miguel.rodriguezgarcia@kaust.edu.sa

R. Valencia-García (✉)
Department of Informatics and Systems Universidad de Murcia, Murcia, Spain
e-mail: valencia@um.es

J. Mejia et al. (eds.), *Trends and Applications in Software Engineering*,
Advances in Intelligent Systems and Computing 405,
DOI 10.1007/978-3-319-26285-7_19

1 Introduction

Smart phones on the Android and iOS platforms enjoy ubiquity previously unseen in computing amongst general consumers with over than 1.4 million apps available on the App Store alone.[1] However, how much information value does it offer developers beyond the aggregated public sentiments that users have towards their app [1]?

Successfully identifying and polarising app aspects that users most discuss and value would focus developer review scrutiny, prioritising their efforts favouring more refined releases. To reach the goal of obtaining information from user reviews for designers and developers, we start by investigating how users express and regard aspects of apps in these opinions.

While related work attempts to model the statistical properties of mobile app re-views [2–5], the content that users communicate in short reviews, is relatively unexplored beyond manual inspection or initial efforts into content analysis [1, 6, 7]. These works [5, 6] studied longer, verbose reviews, affording derivation of a domain corpus, unlike the generally short nature of App Store reviews [1, 3, 7], that skews word weighting during topic extraction. The highly skewed nature of app review size [3] results in succinct reviews resisting the probabilistic techniques of Natural Language Processing (NLP) when analysed holistically. In addition, even if NLP yielded probabilistic tagging of the natural language employed by users, the relationship between jargon and natural language remains disparate [8]. Hence, we develop three ontologies in this work to support content detection in app reviews towards bridging user expressions of app functionality, quality, and the emotional affect the app gives rise to.

In this work, we outline an exploratory ontology development approach to formalise vocabulary used to describe software from the perspective of users. This formalised conceptualisation is framed in the context of software discussion. Prior research has detailed the construction of ontologies in software engineering [9] and the alignment of semantic and knowledge evolution [10]. By referencing known software quality models [11–13] to form the core classifications for our domain ontologies, the segregation of software aspects presented in these models is maintained. Additionally, approach to determine the connotation polarity of words that are semantically relevant to these software aspects in identifying areas of success and improvement is outlined.

This paper is structured as follows: In our Background, we explore relevant work about ontology development. We then outline our ontology development and refinement in our Approach, preceding our Findings and Discussion. Our work is concluded and future efforts in this research area are projected in our Conclusions and Future Work.

---

[1]Apple 2015 1st Quarter Press Release: http://www.apple.com/pr/library/2015/01/08App-Store-Rings-in-2015-with-New-Records.html (last accessed 3rd Sept 2015).

2 Background

An ontology is the "formal and explicit specification of a shared conceptualization" [14], providing a formal structured knowledge representation within a reusable and shareable construct. Ontologies offer a common vocabulary for a domain and with different levels of formality, define semantics of terms and the relationship between them. Although classes in the ontology are usually organized into taxonomies, there are definitions of ontologies less restrictive in the sense that taxonomies are considered to be full ontologies [14].

Domain ontologies are constructs that store relevant knowledge of an application domain and platform, modelling relevant real world aspects. Ontologies are being applied to different domains such as biomedical [15], neuroscience [16], information management [17], finance [18], innovation management [19] or cloud computing [20]. In this work, we explore and develop ontologies framed within the software context, pertaining to emotional, functional and non-functional (henceforth referred to as quality) domains.

On the other hand, semantic annotations bridge the ambiguity between natural language and their computational representation in a formal language through ontologies [21]. This process annotates the document with meta information on ontological elements (attributes, concepts, relationships and instances) associating text fragments to relevant document points or external information. The annotation method may vary between pattern-based (based on discovery rules) and machine learning-based (relying on probabilistic and induction techniques) approaches [20]. However, both approaches allow for documents that can be processed by both humans and automated agents [23]. Existing research classifies annotation approaches as manual, semi-automatic (supported by automatic suggestions) or automatic (based on computer annotation processes). The use of semi-automatic approaches offer effective results within a reasonable time frame [21].

Hence, for this work, we employ a semi-automatic annotation approach based on GATE [24]. As semantic annotations may yield text fragment mappings that require association with overarching concepts and external reference data, the software quality models [11–13] serve as the knowledge base that we reference in the enrichment of the ontologies. The ISO 9126 [13] was selected as the model to reference, as the model provides explicit separation of functional and non-functional software aspects.

In addition to the ISO conceptualization of attributes, we look towards the review dataset studied in [1, 2, 4, 7]. From the primary dataset of 8.7 million app reviews, we extracted all Health & Fitness (H&F) reviews from both Free and Paid price points. The H&F category was selected due to the generally larger review size observed [1, 3, 4]. This enrichment dataset is composed of 330,408 app reviews from 800 apps across all star ratings, and also encompasses the Socrates Dataset [25].

3 Approach

Our ontology construction method uses a semi-automatic semantic annotation approach based on machine-learning [22]. We developed and utilised tooling based on GATE [24], and reference the ISO 9126 [13] quality model due to the explicit separation of functional and non-function software attributes.

Prior to semantic annotation, all raw app reviews were tokenised into separate words. Each word was checked for spelling through the use of Levenshtein distance based libraries such as Jazzy[2] and Lucene[3] for spelling suggestions which were manually performed in the context of the surrounding text.

3.1 Semantic Annotation

Two modules compose the semantic annotation system developed: the extractor module, which builds a list of terms of the ontology classes, and the NLP (Natural Language Processing) module that obtains the semantic annotation for the analysed text fragment in accordance with the domain ontologies.

The semantic annotation process first tokenises each review again to verify the spelling of each word. Each token is then filtered using Lucene's stop word filters. For each word that does not trigger the stop word filter, we analysed it with our binary classifiers.

The binary classifier was developed using LingPipe[4] and is a boolean dynamic language model classifier for use when there are two categories, but training data is only available for one of the categories. This classifier is based on a single language model and cross-entropy thresholding. It defines two categories, accept and reject, with acceptance determined by measuring sample cross-entropy rate in a language model against a threshold.

We use two binary classifiers for this experiment, each taking the same word as an argument and returning a binary result per classifier. Each classifier is fed a list of polarised sentiment words (positive or negative) derived from the word sets developed by Hu and Liu [26]. These classifiers result in either a True or a False flag that indicate the presence or absence of the argument word in the associated polarised word list, respectively. This allows us to draw conclusions regarding the connotations of each word. The conclusions drawn from each result can be seen in Table 1.

Once the connotation of the word is established, it is classified within the domain ontology using semantic annotation. This annotation system is developed based on GATE [22], which includes an information extraction system. This system is a set

[2]http://jazzy.sourceforge.net (last accessed 23/06/2015).

[3]https://lucene.apache.org (last accessed 23/06/2015).

[4]http://alias-i.com/lingpipe (last accessed 23/06/2015).

Table 1 Results from our binary classifiers and how word connotations are concluded

Binary classifier results		
Positive flag	Negative flag	Conclusion
True	True	Word has a mixed connotation
True	False	Word has a positive connotation.
False	False	Word has a neutral connotation
False	True	Word has a negative connotation

of modules comprising a tokeniser, a gazetteer, a sentence splitter, a parts-of-speech tagger, a named entities transducer and a co-reference tagger. Our semantic annotator uses the gazetteer, to build three lists of words as part of each ontology domain. The structure of each list is an enumeration of words involving attributes (emotion, functionality, quality). These attributes form the concept Uniform Resource Identifier (URI), or ontology class, for the word under analysis.

The function of the semantic annotation serves to associate each analysed word from the review to the attribute defined in the ontology domain through the URI of a class on the ontology domain. As a result of the whole process two classifications are obtained; the first classification expresses the connotation of the tokenised word as either positive, negative, neutral or mixed. The second classification expresses the domain related to the analysed word, for example in the quality domain, such as reliability or usability. The union of both classifications represents the semantic analysis of polarity. For instance, the Positive connotation and the Screen attribute of functionality.

Post initial semantic annotation, the information is organised in a hierarchical structure based on the type of relation (IS-A). This is followed by the identification and obtaining of the key words used for the definition of each category within the ontology. The ontology undergoes an enrichment process using semantic information derived from the App Store and WordNet [27]. The App Store was mined for app descriptions on a per category (i.e. Entertainment, Health & Fitness, Social Networking) basis, with each app description tokenised for Term Frequency (TF) analysis. Frequently occurring words within a common App Store category provided supplementary association information Extracts of the Emotional (left), Functional (middle) and Quality for word semantics. In the case of WordNet, relationships such as antonyms and synonyms were extracted to be piped into the respective domain ontology.

Three ontologies were developed: Functionality, Emotional and Quality. In order to build each ontology, different methods were employed. The construction of each ontology is detailed as follows:

1. Emotion Ontology: The development of this ontology involved the introduction of a hierarchical structuring of emotional terms, as shown in the left column of Fig. 1.

Fig. 1 Extracts of the emotional (*left*), functional (*middle*) and quality (*right*) ontologies

The following example scenario outlines the process of construction for the term "pensiveness". Upon tokenisation of "pensiveness", semantically relevant information such as the associated synonyms, antonyms or heteronyms would be parsed and extracted from existing information. Each term that has been mined is then annotated and represented by a class in the ontology. The result of this process is then formalised as a conceptualisation.

2. Functional Ontology: The construction of this ontology commenced with the tokenisation and TF analysis of the app review dataset. The TF results provided probabilistic data pertaining to user expressions in the context of reviews and an order in which to cull the terms by. These results are then culled for terms not relevant to software functionality as described in the software quality model. This culling process was supported by interfaces developed to parse computing dictionaries for definitions, and to verify semantic association and relevance between the text fragment and the known software functionality attributes. The computing dictionaries used for this process were the Computer Desktop Encyclopaedia,[5] the Free On-Line Dictionary of Computing[6] and Gartner.[7]

At this point of construction, the store categorised TF results of the app descriptions were parsed for further semantic annotation of the reviewer text. As

[5]http://www.computerlanguage.com/ (last accessed 23/06/2015).

[6]http://foldoc.org (last accessed 23/06/2015).

[7]http://www.gartner.com/ (last accessed 23/06/2015).

these app descriptions were written by the developers, the probabilistic data from the TF analysis provided a natural language representative resource from which we could infer jargon to natural language relevance. Fragments identified in this phase are processed for semantic annotation, and should relevance be detected, are reflected in the ontology class of "category" (depicted in Fig. 1). Between the app description, software quality model and identified semantic relevance, the text fragments under analysis. Text fragments that did not bear semantic relevance were culled from entry into the domain ontology.

Annotated text fragments that were not culled for irrelevance were manually inspected and grouped amongst two authors independently. Common groupings developed into ontology categories that represented general functional aspects related with computing and software, as depicted in the middle column of Fig. 1. Post categorisation, the synonym, antonym and heteronym relationships are identified and associated to build the class conceptualisation in the ontology.

3. Quality Ontology: Quality (ISO9126+Wordnet): This ontology contains the quality categories describe on ISO9126. These categories have been enriched through the Wordnet repository. This enriching process has been done taking into account semantic relationships such as: synonyms, antonyms and related words. The ISO 9126 quality model is used to annotate the reliability attributes such as: Efficiency (Compliance, time based), Functionality (accuracy, security, interoperability, etc.) and the functional ontology covers the functionality topics such as: categories of apps like game, social, weather, and functionality issues like: audio, hardware, networks. ISO 9126 ontology covers aspects related to reliability of software.

This ontology models the software quality, so that is why a structure was selected. The ISO report 9126 was taken as a hierarchy to build the ontology (See Fig. 1's right column). The quality ontology hierarchy was created with the same hierarchy that ISO describes on its report 9126. The next step was to populate the ontology through the software able to extract synonym, antonym, heteronym relation from a word.

In the information extraction phase, we have parsed the ISO 9126, Boehm, and McCall software quality models, and appended any synonyms and grammar variations related to quality attributes. This process was informed by WordNet, and performed in a manner similar to ontology construction.

These lists of quality-type words were used as a reference point for the subsequent phase, henceforth referred to as quality aspects. We tokenised and parsed the review text to look for any words that matched a quality aspect. Instances of such words then served as the root for unigram extraction with a phrase length of five. In the phrases extracted from the root, we parsed to detect sentiment words. In the instance that a sentiment word was found, we extracted all phrases that encapsulated both the quality aspect and sentiment.

Once this phase has been completed across the experiment dataset, we grouped the quality aspects that we detected into their software quality attributes. This was first performed on a low level of granularity, by sorting the phrases by content into functional and non-functional types. From there, we manually sorted them into their sub types in accordance with ISO 9126. An example of this is to extract and group all phrases pertaining to Reliability, Portability, and so on, respectively.

For each type of quality aspect, we determine if users convey positive or negative sentiments toward it. The resulting sentiment distribution is also contrasted against the general sentiment of other quality aspects. Based on the most discussed quality aspects, we draft a continuum to plot reviews on, to visualise the type of aspect and the weight of it, informed by the frequency of users raising it. This is done toward identifying if reviews exhibit generalisable properties in sentimental quality aspects.

In closing, once the ontologies were built, the next step was to build the list of terms per each ontology. These list are used by GATE [20] to semantic annotation process.

3.2 Limitations

Only reviews composed of up to five words from a single category were collected for this study. Hence, we are currently unable to generalise our findings beyond reviews of this size and from this category. In addition, the primary dataset comprises reviews from the top ranked apps. Our approach in this paper did not separate hyphenated words for word counting. However, a random selection of 50 reviews shows that this limitation affected less than 2 % of the data. Fake user generated reviews were not addressed as the focus of this study was on information content and value.

The classifier is based on statistical model which is trained with a incomplete list of positive and negative words, and we do not have the neutral and mixed list of words which represents the others categories (positive, negative, mixed and neutral), so if we had those lists the results of the analysis could be improved because in this case the filter could be more accurate.

The ontologies do not represent the domain completely so that is why we can find several mistakes on the annotation process. So the content could be improved in order to obtain classification more precise.

4 Findings and Discussion

Using existing dictionaries and our trained auxiliary dictionary, our system is able to successfully detect and classify 99.6 % of the content in our summarised dataset. The undetected words were considered as spam or noise. We also observed that 27,685 (94.9 %) of the dataset consistent in content and rating to communicate Positive satisfaction, which may be a side effect of the data being collected from successful apps.

Following Positive-type classification, reviews classified as Neutral amounted to 2.2 % of the total reviews processed. Negative and Spam classified reviews both accounted for 1.1 %, while Inconsistent Positive and Inconsistent Negative were only 0.5 and 0.3 % of the dataset, respectively. Interestingly, 8 out of 25 apps have either zero Inconsistent Positive reviews or Inconsistent Negative reviews.

My Baby Today achieved the highest consistent satisfaction amongst users, with 919 (98 %) out of 936 processed reviews classified as consistently positive. Nike +Running presented the next to lowest Positive-type (78.9 %) reviews, while possessing the highest proportions of both Neutral– (5.5 %) and Spam-type (2.9 %) reviews.

In order to achieve the classification, our approach requires as inputs a set of reviews, lists of positive, negative and neutral words and the emotion, functionality and quality ontologies. Some of the metrics of the developed ontology are shown in Table 2.

During a first stage, each review is splitted in separate words. Then each word is individually checked by the spell corrector and filtered by a stop word filter. After that, words are annotated by the semantic annotation module in order to classify them in terms of emotions, functionalities or quality features. Finally, the two-level filter categorises each word depending whether it has positive, negative or neutral polarity.

The final results of the experiment are shown in Table 3. The system obtained the best scores for queries of Paid Apps, with a precision of 0.80, a recall of 0.69, and a F1 measure of 0.74. In general, the system obtains better results in precision (80 % on average) than recall (73 % on average). Hence, these results are promising.

Table 2 Properties of the ontologies

Properties			
	Emotion ontology	Functionality ontology	Quality ontology
Classes	336	55	31
Labels	5019	1147	1224

Table 3 Precision, Recall and F1 of the experiment

Results							
		Annotation process			Connotation process		
Types	Number of reviews	Precision	Recall	F1	Precision	Recall	F1
Free apps	100	0.79	0.68	0.73	0.75	0.70	0.72
Paid apps	100	0.80	0.69	0.74	0.77	0.72	0.75

5 Conclusions and Future Work

In this work, a new approach for analysing mobile app reviews to determine if the content and rating values are consistent, is presented. These reviews offer developers a view of user satisfaction of both their products and an aggregate level of user expectations for mobile apps.

Our experiment to detect if users are consistent in communicating their satisfaction shows promising results. Firstly, short reviews (up to five words) are primarily composed of sentiment words. In addition, we report strong correlation in the text content and numerical rating of these short reviews, indicating minimal information loss if developers or potential users were to only regard the aggregated rating of an app. In particular, we show that any detailed analysis of the content can ignore nearly a quarter of the reviews since they do not contain information beyond the numeric rating.

Our work is focused on short reviews (up to five words), and within a single category. The next steps are to expand our method to longer reviews and to consider other categories (for instance, Games). In particular, we intend to identify the range in terms of word count when reviews are likely to start offering useful and valuable information beyond that captured by the numeric rating.

Acknowledgments This work has been supported by the Spanish private foundation Fundación Cultural Privada Esteban Romero through its research stays grants.

References

1. Hoon, L., Vasa, R., Martino, G.Y., Schneider, J.G., Mouzakis, K.: Awesome! conveying satisfaction on the app store. In: 25th Australian Computer-Human Interaction Conference (2013)
2. Harman, M., Jia, Y., & Zhang, Y.: App store mining and analysis: MSR for app stores. In: 9th IEEE Working Conference on Mining Software Repositories, pp. 108–111. IEEE Press (2012)
3. Vasa, R., Hoon, L., Mouzakis, K., Noguchi, A.: A preliminary analysis of mobile app user reviews. In: 24th Australian Computer-Human Interaction Conference, pp. 241–244 (2012)
4. Hoon, L., Vasa, R., Schneider, J.G., Grundy, J.: An Analysis of the Mobile App Review Landscape: Trends and Implications. Faculty of Information and Communication Technologies, Swinburne University of Technology, Melbourne, Australia, Tech. Rep. http://hdl.handle.net/1959.3/352848
5. Iacob, C., Harrison R.: Retrieving and analyzing mobile apps feature requests from online reviews. In: 2013 10th IEEE Working Conference on Mining Software Repositories (MSR), pp. 41–44. IEEE (2013)
6. Platzer, E.: Opportunities of automated motive-based user review analysis in the context of mobile app acceptance. In: 22nd Central European Conference on Information and Intelligent Systems. CECIIS '11, pp. 309–316 (2011)
7. Hoon, L., Vasa, R., Schneider, J.G., Mouzakis, K.: A Preliminary Analysis of Vocabulary in Mobile App User Reviews. In: 24th Australian Computer-Human Interaction Conference, pp. 245–248 (2012)

8. Binkley, D., Hearn, M., Lawrie, D.: Improving identifier informativeness using part of speech information. In: 8th Working Conference on Mining Software Repositories, MSR '11, pp. 203–206. ACM, New York (2011)
9. De Nicola, A., Missikoff, M., Navigli, R.: A software engineering approach to ontology building. Inf. Syst. **34**(2), 258–275 (2009)
10. Li, Y.F., Zhang, H.: Integrating software engineering data using semantic web technologies. In: 8th Working Conference on Mining Software Repositories. MSR '11, pp. 211–214. ACM, New York (2011)
11. McCall, J.A., Richards, P. K., Walters, G.F.: Factors in Software Quality. General Electric, National Technical Information Service (1977)
12. Boehm, B.W., Brown, J.R., Kaspar, H., Lipow, M., MacLeod, G.J., Merrit, M.J.: Characteristics of software quality. North-Holland Publishing Company, vol. 1 (1978)
13. ISO/IEC 9126-4 Software Engineering—Product Quality—Part 4: Quality In Use Metrics, ISO/IEC, Tech. Rep (2002)
14. Studer, R., Benjamins, V.R., Fensel, D.: Knowledge engineering: principles and methods. Data Knowl. Eng. **25**(1), 161–197 (1998)
15. Ruiz-Martínez, J.M., Valencia-García, R., Martínez-Béjar, R., Hoffmann, A.: BioOntoVerb: a top level ontology based framework to populate biomedical ontologies from texts. Knowl. Based Syst. **36**, 68–80 (2012)
16. Prieto-González, L., Stantchev, V., Colomo-Palacios, R.: Applications of ontologies in knowledge representation of human perception. Int. J. Metadata Semant. Ontol. **9**(1), 74–80 (2014)
17. Colomo-Palacios, R., Garcia-Crespo, A., Soto-Acosta, P., Ruano-Mayoral, M., Jiménez-López, D.: A case analysis of semantic technologies for R&D intermediation information management'. Int. J. Inf. Manage. **30**(5), 465–469 (2010)
18. Lupiani-Ruiz, E., García-Manotas, I., Valencia-García, R., García-Sánchez, F., Castellanos-Nieves, D., Fernández-Breis, J.T., Camón-Herrero, J.B.: Financial news semantic search engine. Expert Syst. Appl. **38**(12), 15565–15572 (2011)
19. Hernández-González, Y., García-Moreno, C., Rodríguez-García, M.Á., Valencia-García, R., García-Sánchez, F.: A semantic-based platform for R&D project funding management. Comput. Ind. **65**(5), 850–861 (2014)
20. Rodríguez-García, M.Á., Valencia-García, R., García-Sánchez, F., Samper-Zapater, J.J.: Ontology-based annotation and retrieval of services in the cloud. Knowl. Based Syst. **56**, 15–25 (2014)
21. Oren, E., Moller, K., Scerri, S., Handschuh, S., Sintek, M.: What are semantic annotations. Relatório técnico, DERI, Galway (2006)
22. Reeve, L., Han, H.: Survey of semantic annotation platforms. In: 2005 ACM symposium on Applied computing, pp. 1634–1638. ACM (2005)
23. Kiyavitskaya, N., Zeni, N., Cordy, J.R. Mich, L., Mylopoulos, J.: Semi-automatic semantic annotations for web documents. In: SWAP (2005)
24. Cunningham, H. Tablan, V. Roberts, A., Bontcheva, K.: Getting more out of biomedical documents with gate's full lifecycle open source text analytics. PLoS computational biology, vol. 9, no. 2 (2013)
25. Mouzakis, K.: Hoon, L., Rajesh, V.: Socrates Mobile App Review Dataset (2013)
26. Hu, M., & Liu, B. Mining and summarizing customer reviews. In: 10th ACM SIGKDD International Conference on Knowledge Discovery and Data Mining. pp. 168–177. ACM (2004)
27. Fellbaum, C.: WordNet. Wiley Online Library (1999)

A Computational Measure of Saliency of the Shape of 3D Objects

Graciela Lara, Angélica De Antonio and Adriana Peña

Abstract The shape of an object is a basic characteristic that when attracts the viewers' attention represents a salient feature. In this paper we propose a computational measure of saliency of the shape of 3D objects in virtual reality, based on the proportion of empty and full space within its bounding box. This measure of saliency is part of a computational model aimed to the selection of appropriate reference objects to facilitate the location of objects within a 3D virtual environment. An experiment was conducted to understand to which extent the proposed measure of saliency matches with the people's subjective perception of saliency; results showed a good performance of the metric.

Keywords 3D objects · Shape saliency · Voxelization

1 Introduction

The visual attraction of an object is largely determined by its basic features such as color, size or shape. Regarding specifically on the shape of a 3D object, it can be defined in a verbal or a graphic way. The shape of a 3D object represents its exterior geometrical figure; when described, we can use some of these characteristics, such as: area, perimeter, diameter, minimum and maximum distance from the center of

G. Lara (✉) · A. Peña
CUCEI of the Universidad de Guadalajara, Av. Revolución 1500, Col. Olímpica, 44430 Guadalajara, Jalisco, Mexico
e-mail: graciela.lara@red.cucei.udg.mx

A. Peña
e-mail: adriana.pena@cucei.udg.mx

A. De Antonio
Escuela Técnica Superior de Ingenieros Informático of the Universidad Politécnica de Madrid, Campus de Montegancedo, 28660 Boadilla del Monte, Spain
e-mail: angelica@fi.upm.es

J. Mejia et al. (eds.), *Trends and Applications in Software Engineering*, Advances in Intelligent Systems and Computing 405,
DOI 10.1007/978-3-319-26285-7_20

mass, axes, angles, and its surrounding space (i.e. bounding box). The visual attraction of these characteristics is linked to the concept of saliency.

The concept of the saliency has been applied by psychologists and computer vision researchers mainly to facilitate the computer recognition of objects and to locate their salient areas in images (e.g. [1–6]). According to Kapur [7], visual saliency is the process of the association of objects and their representation that attracts attention and captures people's thinking and behavior. While the ability of the human visual system to detect the visual saliency of an object, or a set of objects, is extraordinarily fast and reliable, the computational modeling of this basic intelligent human behavior still remains as a challenge [1]. In recent decades the study of the visual saliency of the shape of an object has attracted the researchers' attention. A good approach to this subject conducted by Lee et al. [8], introduced the concept of salient mesh, defined in a scale-dependent manner using a center-surround operator on Gaussian-weighted mean curvature. They calculated salient regions of a 3D mesh using its geometry at different scales; the final saliency measure is the aggregation of these scales with a non-linear normalization.

A quantitative measure of saliency proposed by Mitra et al. [9], is based on the extraction of significant Euclidean symmetries at all scales. With the extracted graph of symmetry representation, important high-level information about the structure of a geometric model can be captured, which represents the shape of the object.

Gal and Cohen-Or [10] introduced a method for partial matching of represented surfaces by triangular meshes based on the theory of saliency of visual parts. They considered the saliency of a part, as a function of its size relative to the whole object, the degree to which it protrudes, and the strength of its boundaries.

Yamauchi et al. [11] proposed an approach using a uniform set of viewpoints connected with the closest one, to form a spherical graph where each edge is weighted by a similarity measure between the two views from its incident vertices. The similar views are gotten using a graph partition process, where the centroids are considered the most representative views. This measure orders representative views from the most salient to the least one. The saliency is measured by the mean curvature of a 3D mesh model.

Other methodology proposed by Castellani et al. [12] is based on the detection of salient points in several views of an object. They evaluated the saliency by using the principles of visual saliency of 3D meshes; then a statistical learning approach is applied to describe the salient points from different views and a Hidden Markov Model (HMM) is used to model each salient point.

Under a technique called extremum lines, Miao and Feng [13] presented a perceptual saliency measure that captures the features of the visual saliency of the shape, yielding visually pleasant results. This measure is based on extreme lines considered as the ridge-valley along the principal curvature directions on triangular meshes. Latter et al. [14] presented other method using salient regions in a polygonal mesh of a 3D shape. This approach considers the illumination and shading to enhance the geometric salient features of a 3D model. The calculation of this measure uses only the surface normals in the context of visual attention-driven shape depiction. This estimation has the limitation of enhancing shape depiction caused by the variation of the diffuse lighting component.

Finally, Tao et al. [15] proposed a method that considers salient regions of a mesh, using the detection of the manifold ranking in a descriptor of Zernike coefficients. They used the regions of the mesh that are most scattered in the spatial domain with no regard of their locations and cardinality.

These measure saliency approaches were designed and evaluated from different perspectives, achieving quite good results. However, none of them are applied directly to the 3D object; in all aforementioned cases the 3D model is decomposed into a set of 2D projections, maps or views. In contrast, our proposal uses the whole shape of the object, without the need of transforming the 3D model in any way. Furthermore, our measure approach can be used in 3D objects without regard of their form (regular or irregular), position, scale or orientation. And because it does not require a complex preprocessing, its computational cost is minimal. On the other hand, our proposal for measuring saliency is oriented to answer the question of what is the most salient object in a set of 3D objects, and it is characterized by its promptness, simplicity and efficiency. Although, despite its simplicity, empirical results showed that the most salient objects given by our metric matches to the most salient objects according to human perception. In the next section, our measure of saliency is described in detail.

2 A Measure of Saliency of the Shape

The measure is based on the central idea of the volume of a pre-voxelized object. To apply this measure of saliency to the shape of the 3D object, it is necessary to consider the following three previous steps:

(1) The object requires first being voxelized. The process of voxelization is the segmentation of an object into small cubic portions, a unit called voxel, which conform and represent the three dimensional object, as shown in Fig. 1. The voxelization process characterizes both the size and the shape of an object. Regarding the size of the voxel (vsi), in virtual environments is common to work with an approximation to the size in terms of centimeters; the size of the voxel is here treated as one cubic centimeter.

(2) Then, the total number of voxels (nv) for the object is obtained through the voxelization algorithm by counting each voxel, whenever it detects a correct one.

(3) Then, the number of voxels (nv) is multiplied by the voxel size (vsi) to obtain the object volume (vo) in cubic centimeters as in Eq. (1).

$$vo = nv * vsi \tag{1}$$

Once calculated measures regarding the volume of the 3D object, we assumed that the flatter is the surface of an object, the less salient it is, and inversely, objects

Fig. 1 View of two voxelized 3D objects

with high pointedness tend to be perceptually more salient. With this in mind, we propose the following method for a fast estimation of shape saliency: *calculating the proportion of empty space and full space in voxels in the bounding box of each object*. For this, first the volume (size) of the bounding box in voxels (siBBox) is obtained through the length, width and height of the box and the size of the voxel. A bounding box is a three-dimensional box that delineates the boundaries of an object (see Fig. 2, the bounding box is represented with discontinued lines). Then the volume in voxels of the object (vo) is subtracted from this number. The result of this operation is called "Empty space" (Es), because it represents the number of free voxels within the bounding box that are not part of the object, Eq. (2).

$$Es = siBBox - vo \tag{2}$$

The volume of the object in voxels corresponds to the "Full space" (Fs) inside the bounding box. Figure 2 illustrates the empty space of a 3D object within its bounding box.

Finally, the empty space (Es) is divided by the total number of voxels in the bounding box (siBBox), giving the proportion of empty space, Eq. 3.

$$Ssh = Es/siBBox \tag{3}$$

This provides a rough estimation of the saliency of an object's shape. However, as our final goal is to be able to select the most salient 3D objects within a 3D virtual environment, and considering that saliency is strongly dependent on the context in which objects are located, we apply a normalization process over the

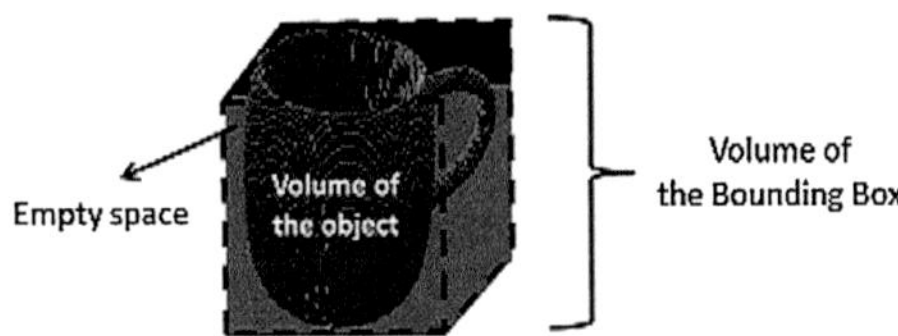

Fig. 2 Empty and full space of the bounding box of a 3D object

previously computed value. The object with the highest proportion of empty space will have the greater saliency value '1' and the rest a weighted value in proportion to the greater value. Normalized saliency by shape is represented by $Ssh_{[0\text{-}1]}$.

3 Experimental Evaluation of the Shape Saliency Metric

With the aim to verify our hypothesis: that people perception of saliency is similar with our measure of saliency approach, we designed and carried out the next described experiment.

3.1 Method

Participants. Forty undergraduate students of the Escuela Técnica Superior de Ingenieros Informáticos of the Universidad Politécnica de Madrid, thirty male and ten female, with ages in the range of 18 to 25 years, voluntarily participated.

Materials, devices and situation. The experiment was carried out in a laboratory with suitable lighting condition. Each participant's session was run in a SONY laptop computer, model VGN-CS270T, with a processor Intel ® Core (TM) 2 Duo CPU P8600, 2.40 GHz, 4.00 GB memory, using a mouse and a keyboard. A computational application was developed to implement and test the metric using the Unity 3DTM platform, with some scripts created in C# programming language. Results were automatically recorded in a .csv (comma-separated values) file and afterwards statistically analyzed using the SPSSTM (Statistical Product and Service Solutions) application.

The voxelization process of the 3D objects was made through a master script of voxelization of Unity 3DTM adjusted to extract the number of voxels and so calculating the volume of each object.

Design and procedure. The designed experimental system consisted of 25 scenarios (also called trials), each presenting four 3D objects. One hundred 3D objects were extracted randomly from the Princeton Shape Benchmark (PSB), which contains 1,814 3D models and is public available in the World Wide Web [16]. The set of the selected objects for this experiment were adjusted to have the same size scale. Each of the objects was voxelized and the numbers of voxels was counted, the time for the voxelization process of each object varied from 0.405 to 1.097 s.

Participants were informed that their task during the experiment was to place the four objects provided on each trial on an empty platform in front of them (see Fig. 3), ordering them from left to right according to their shape saliency, this concept was explained to them as "the capability of the object's shape to attract their attention". Therefore, the most striking object by its shape should be placed to the left. Participants were given a brief demonstration of the system on how to place

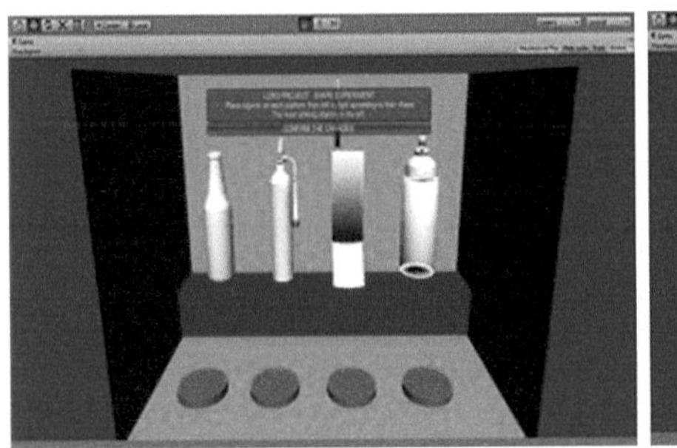
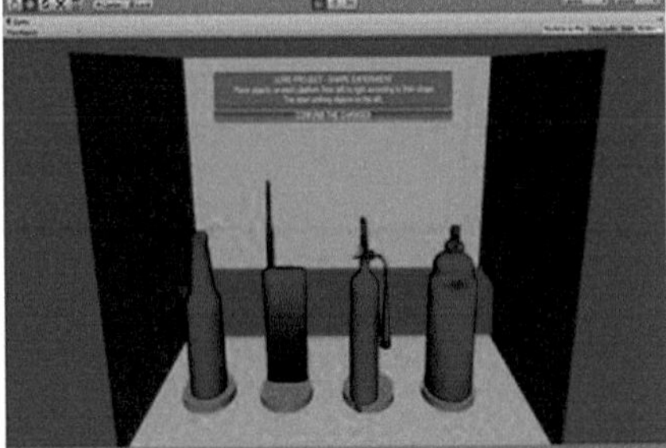

Fig. 3 View of an experimental system trial

each object on the platform as shown in Fig. 3. Also, we explained to participants that they could make all necessary place changes, before they confirmed the final order of objects for each trial. Participants were asked to provide basic personal information as their, age and gender, within the system. Each person lasted about 8 min to complete the 25 trials.

3.2 Statistical Analysis of Results

We conducted several statistical analyses in order to evaluate our general hypothesis of how similar is our metric to the perception of saliency of the participants. We analyzed the following three aspects:

1. *The choice of the first or second object as the most salient.* Because the salience of the shape of an object is a very subjective characteristic, and given that there is no precise order with which to compare the given order of our metric and each of the given orders of the participants, the comparison required to be flexible enough to admit variations, but at the same time capable to give information about the performance of the metric. We decided to measure the extent to which the first or second most salient objects, according to our metric of saliency, matched the object placed by the participants in the first position. This condition helped us to identify if the most salient objects to the human visual system corresponds with high-valued objects by our metric.
2. *Simple error: comparison of the mode with respect to the metric.* In this second statistical analysis, we compared for each trial the order given by our metric (Table 1 '(a)') and the order given by the statistical mode (Table 1 '(b)'). The mode of the most voted object in each position for each trial was obtained, counting the number of times that an object was placed in each of the four positions within each trial. Then, we computed the distance between each pair of objects (the one given by the metric and the most voted one) in each position of each trial, according to the value assigned to each object by our metric.

Table 1 Order of the objects in each trial: (a) based on our metric of saliency and (b) based on the mode. (When the order is the same in both, the cell is highlighted in green)

		(a)		
Trial	**Pos_1**	**Pos_2**	**Pos_3**	**Pos_4**
Trial_1	Lamp	Sextant	Antenna	**Dice**
Trial_2	ShaveMac	Apollo13	**Lollypop**	**Disket**
Trial_3	**Mallet**	Apple	**Shovel**	Door
Trial_4	Sofa	**MayaPira**	**Arrow**	Drum
Trial_5	**Spider**	**Microsco**	**Axe**	**Drumbell**
Trial_6	**Enterprise**	**MobilePh**	Bat	SteakKni
Trial_7	**MobilePh**	Extinguis	BeerBottl	Spray
Trial_8	Bicycle	F16Plane	Motorcyc	**StreetLan**
Trial_9	Submarine	PailCube	Binocular	FaxMach
Trial_10	Fence	**PanelScre**	Table	Biotank
Trial_11	**FerrariF3**	**Pear**	**TeaCup**	**BitDrill**
Trial_12	**BoingPlan**	Pencil	FishTank	Televisio
Trial_13	TeremeoL	**Piano**	Book	FlashLigh
Trial_14	**Flute**	PicnicBe	TheetBru	**Briefcase**
Trial_15	PipeSmok	TieFighte	**C64Chip**	**GarbageC**
Trial_16	**Positional**	ToyBear	GasMark	**Camera**
Trial_17	Glasses	**Candle**	PsxContr	**Train**
Trial_18	Hammock	Tricycle	CandyPo	Revolver
Trial_19	Handbell	**Truck**	CanpyBe	**Dice**
Trial_20	Hat	Umbrella	Roadster	CashRegi
Trial_21	Headset	Unicycle	**RocketM**	CasketBo
Trial_22	**Hind24H-**	Chair	Violin	Rollerball
Trial_23	**Rose**	CleaverK	Wagon	Hourglass
Trial_24	**Hydrant**	Washing	CPU	**RugbyBal**
Trial_25	SaberSwo	X-Wing	Keyboard	**Cross**
		(b)		
Trial	**Pos_1**	**Pos_2**	**Pos_3**	**Pos_4**
Trial_1	Sextant	Antenna	Lamp	**Dice**
Trial_2	Apollo13	ShaveMac	**Lollypop**	**Disket**
Trial_3	**Mallet**	Door	**Shovel**	Apple
Trial_4	Drum	**MayaPira**	**Arrow**	Sofa
Trial_5	**Spider**	**Microsco**	**Axe**	**Drumbell**
Trial_6	**Enterprise**	**MobilePh**	SteakKni	Bat
Trial_7	**MobilePh**	Spray	Extinguis	BeerBottl
Trial_8	F16Plane	Motorcyc	Bicycle	**StreetLan**
Trial_9	Binocular	Submarine	FaxMach	PailCube
Trial_10	Biotank	**PanelScre**	Fence	Table
Trial_11	**FerrariF3**	**Pear**	**TeaCup**	**BitDrill**
Trial_12	**BoingPlan**	Televisio	Pencil	FishTank
Trial_13	FlashLigh	**Piano**	TeremeoL	Book

(continued)

Table 1 (continued)

		(a)		
Trial_14	**Flute**	TheetBru	PicnicBe	**Briefcase**
Trial_15	TieFighte	PipeSmok	**C64Chip**	**GarbageC**
Trial_16	**Positional**	GasMark	ToyBear	**Camera**
Trial_17	PaxContr	**Candle**	Glasses	**Train**
Trial_18	Tricycle	Revolver	Hammock	CandyPo
Trial_19	CanpyBe	**Truck**	Handbell	**Dice**
Trial_20	CashRegi	Roadster	Hat	Umbrella
Trial_21	Unicycle	CasketBo	**RocketM**	Headset
Trial_22	**Hind24H-**	Violin	Rollerball	Chair
Trial_23	**Rose**	Wagon	Hourglass	CleaverK
Trial_24	**Hydrant**	CPU	Washing	**RugbyBal**
Trial_25	X-Wing	Keyboard	SaberSwo	**Cross**

Finally, to get the value of the simple error (SE) we added the distances in saliency for all positions of each object of each trial. We propose the simple error as a measure of the distance between the order provided by the saliency metric and the most general opinion of participants.

In order to further evaluate the validity of our metric, we computed two additional variables: (EM) the error in movements (or jumps) when comparing the order given by each participant in each trial, with respect to the order given by our metric; and (ES) the error in saliency when comparing the order given by each participant in each trial, with respect to the order given by our metric.

Based on these variables we made two correlation analysis: (1) between the value of the simple error (SE) of each trial and the standard deviation of error in movements (EM) for all participants in each trial; and (2) between the value of the simple error of each trial and the standard deviation of the error in saliency (ES) for all participants in each trial. We wanted to explore if higher diversity between participants leads to higher simple error, in other words, if we can expect that trials in which the distance between the participant's and the metric's order were more variable, were also the trials in which the order generated by our metric was less prototypical.

3. *Comparison between the simple error and the possible diversity in saliency of all trials.* In a last statistical analysis we sought to evaluate a new hypothesis: the greater the diversity of the trial in saliency, the lower the simple error. We expected participants to adhere more closely to the reference order provided by the metric when there were big differences in saliency between the objects presented in the trial. To test this hypothesis we calculated the correlation between these two variables. Results are discussed in the following section.

4 Results and Discussion

The results of the first statistical analysis indicate that our metric effectively predicts the objects that humans tend to perceive as the most salient. The number of possible orders for the 4 objects of each trial is 24, but only half of them fit with the restriction established in our first statistical analysis, that is, to include the first or second most salient objects in the first position. Assuming random orders provided by participants for each trial, we would expect that 50 % of the orders would fit our condition. However, by the actual order given by each of the 40 participants in each of the 25 trials, we got a mean of 16.5 matches for participant. These 16.5 matches represent a 66 % of all possible matches, which exceeds the expected 50 % value by random success. In such a way that our metric is not random and it is indeed measuring what it is expected to measure.

The results in the second statistical analysis showed a degree of 37 % agreements between the order based on the mode and the reference order given by the metric. This percentage demonstrates that our metric is a good representative of the saliency perception in a prototypical person.

We also employed Pearson's correlation coefficients as indicators of the relationship between (1) the simple error and the standard deviation of the error in movements, and (2) the simple error and the standard deviation of the error in saliency. According to these coefficients, both correlations are positive, but low. The sets of points in the scatter diagrams of the Fig. 4a, b allow us to appreciate the linear relationships of both correlations. In the case of Fig. 4a the correlation index was ($r = 0.14$), and in Fig. 4b of ($r = 0.27$).

For the third statistical analysis we also used Pearson's correlation coefficients. The result showed a value of correlation ($r = 0.28$), with a positive trend (see Fig. 5). This indicates that the relationship between variables is positive and low, and therefore the greater the diversity of the trial in saliency, the higher the value in

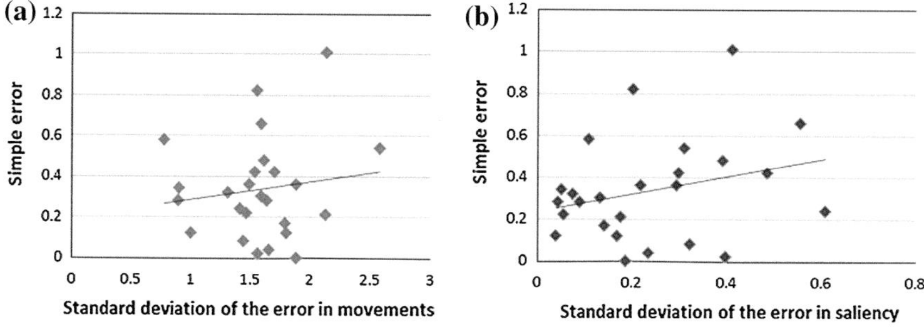

Fig. 4 Dispersion diagrams showing the correlation between **a** the simple error and the standard deviation of the error in movements; **b** the simple error and the standard deviation of the error in saliency

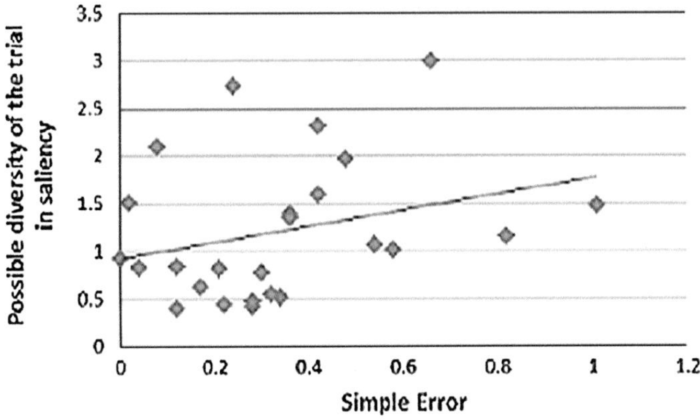

Fig. 5 Dispersion diagram showing the correlation between the simple error and the possible diversity in saliency

the simple error. Here, we must reject the third aspect of our hypothesis in which we assumed that when the difference in saliency between the objects of one trial is high, it would be easier to find agreement between the order of participants and the order generated by the metric, but this is not happening.

5 Conclusions

We proposed a quick and simple approach to measure the saliency for the shape of 3D objects, experimental results demonstrated that our metric is efficient and fits to an extent with the perception of saliency of the participants. Furthermore, our algorithm is easy to implement, requires minimal space and computational time. On the other hand, our metric of saliency can be applied to all kinds of objects with regular and irregular geometry and it is adaptable to all scales of the 3D models. We found that our metric is comparable to a large extent with the perceptual saliency of human beings. Future works should explore other aspects contributing to the visual saliency of 3D objects other than their shape, like color or size, and it also should take into consideration the influence of the context in which an object is placed, for a complete model of saliency for 3D objects.

Acknowledgments Graciela Lara holds a PROMEP scholarship in partnership with the UDG (UDG-685), Mexico. We also thank the students Adrián Calle Murillo, Roberto Mendoza Vasquez, and Álvaro Iturmendi Muñoz for their help in the implementation of the metric and the experimental software application and materials.

References

1. Hou, X., Zhang, L.: Saliency detection: a spectral residual approach. In: Computer Vision and Pattern Recognition 2007, pp. 1–8. IEEE Minneapolis MN (2007)
2. Itti, L.: Quantitative Modelling of Perceptual Salience at Human Eye Position, vol. 14 (4/5/6/7/8), pp. 959–984. Taylor & Francis Group Psychology Press Visual Cognition (2006)
3. Itti, L., Koch, C., Niebur, E.: A model of saliency-based visual attention for rapid scence analysis. IEEE Trans Pattern Anal. Mach. Intell. **20**(11), 1254–1259 (1998) (IEEE)
4. Li, J., Levine, M.D., An, X., Xu, X., He, H.: Visual saliency based on scale-space analysis in the frequency domain. IEEE Trans. Pattern Anal. Mach. Intell. **35**(4), 996–1010 (2013) (IEEE)
5. Raubal, M., Winter, S.: Enriching wayfinding instructions with local landmarks. In: 2002 Proceedings Second International Conference GIScience, pp. 243–259. Springer Berlin Heidelberg, Boulder, CO, USA (2002)
6. Sampedro, M.J., Blanco, M., Ponte, D., Leirós, L.I.: Saliencia Perceptiva y Atención. In: La Atención (VI) Un enfoque pluridisciplinar. Edited by Añaños E, Estaún, S., Teresa MM. España: pp. 91–103, Universidad Autonoma de Barcelona (2010)
7. Kapur, S.: Psychosis as a state of aberrant salience: a framework linking biology, phenomenology, and pharmacology in schizophrenia. Am. J. Psychiatry **160**(1), 91–103 (2003)
8. Lee, C.H., Varshney, A., Jacobs D.W.: Mesh saliency. In: ACM Transactions on Graphics (TOG)—Proceedings of ACM SIGGRAPH, pp. 659–666. ACM New York, USA (2005)
9. Mitra, N.J., Guibas, L.J., Pauly, M.: Partial and approximate symmetry detection for 3D geometry. In; ACM Transactions on Graphics (TOG)—Proceedings of ACM SIGGRAPH, vol. 25(3), pp. 560–568. ACM (2006)
10. Gal, R., Cohen-Or, D.: Salient geometric features for partial shape matching and similarity. ACM Trans. Graphics **25**(1), 130–150 (2006) (ACM)
11. Yamauchi, H., Saleem, W., Yoshizawa, S., Karni, Z., Belyaev, A., Seidel, H-P.: Towards stable and salient multi-view representation of 3D shapes. In: IEEE International Conference on Shape Modeling and Applications (SMI'06): 2006, pp. 265–270. IEEE Matsushima (2006)
12. Castellani, U., Cristani, M., Fantoni, S., Murino, V.: Sparse points matching by combining 3D mesh saliency with statistical descriptors. Eurographics **27**(2), 643–652 (2008)
13. Miao, Y., Feng, J.: Perceptual-saliency extremum lines for 3D shape illustration, Springer The Visual Computer, vol. 26(6–8), pp. 433–443. Springer, Berlin (2010)
14. Miao, Y., Feng. J., Pajarola, R.: Visual saliency guided normal enhancement technique for 3D shape depiction. Elsevier Computers & Graphics, vol. 35, pp. 706–712. Elsevier (2011)
15. Tao, P., Cao, J., Li, S., Liu, X,, Liu, L.: Mesh saliency via a ranking unsalient patches in a descriptor space. Elsevier Computer & Graphics, vol. 46, pp. 264–274. Elsevier (2015)
16. Shilane, P., Min, P., Kazhdan, M., Furkhouser, T.: The Princeton Shape Benchmark. In: IEEE Proceedings of the Shape Modeling International 2004, pp. 167–168. IEEE Washington, DC, USA (2004)

Part IV
Information and Communication Technologies

An Open Cloud-Based Platform for Multi-device Educational Software Generation

Raquel Vásquez-Ramírez, Maritza Bustos-Lopez, Agustín Job Hernández Montes, Giner Alor-Hernández and Cuauhtemoc Sanchez-Ramirez

Abstract Nowadays, information technologies play an important role in education. For instance, mobile applications can be used through tablets and smartphones in classrooms, but not limited to this place. Smartphones are some of the most used devices across the world, and represent one of the most cheapest and user-friendly electronic devices for activities such as education, personal administration, communication and leisure, to mention but a few. In this work, we present a cloud-based platform for multi-device educational software generation (smartphones, tablets, web, android-based TV boxes, and smart TV devices). It is noteworthy mention that an open cloud-based platform allows to teachers create their own multi-device software by using a personal computer with Internet access. The goal of this platform is to provide a software tool that allows to teachers upload their electronic contents and package them in the desired setup file for one of the supported devices and operating systems by using our cloud-based platform.

Keywords Education · Multi-device software · Cloud-based platform · Software generation

R. Vásquez-Ramírez (✉) · M. Bustos-Lopez · A.J.H. Montes · G. Alor-Hernández · C. Sanchez-Ramirez
Division of Research and Postgraduate Studies, Instituto Tecnológico de Orizaba, Avenida Oriente 9 no. 852 Col. Emiliano Zapata, Orizaba, VER, México
e-mail: vz.rmz.raquel@gmail.com

M. Bustos-Lopez
e-mail: maritbustos@gmail.com

A.J.H. Montes
e-mail: agus_job@hotmail.com

G. Alor-Hernández
e-mail: galor@itorizaba.edu.mx

C. Sanchez-Ramirez
e-mail: csanchez@itorizaba.edu.mx

J. Mejia et al. (eds.), *Trends and Applications in Software Engineering*, Advances in Intelligent Systems and Computing 405,
DOI 10.1007/978-3-319-26285-7_21

1 Introduction

Information and communication technologies (ICT's) produce meaningful changes in any society. They are able to bring a great amount of information to the remotest places and they represent a powerful tool to reach autonomous, meaningful, and quality learning [1]. The fast growth of ICT's has allowed an increase in the development of Desktop-based, Web-based, Mobile and TV-based applications. Each one of these proposals represents new alternatives to be used as teaching materials to complement classes inside the classroom. Taking this into consideration, in the last years, information technologies emerge as a valuable resource to support the teaching-learning process across all education levels. Nowadays, students learn new knowledge effectively through new technologies. This is due to the familiarity of new generations with emergent technologies. Software applications for mobile devices can be the most significant example of how information technologies can be useful and suitable for students. However, two issues mainly appear. First, some of these educational contents are not those that a professor needs since information is restricted. In contrast with the open educational contents that allows open access to high quality educational resources for software customization [2]. Second, the development of multi-device applications requires an integrative approach involving the support of several specialists in education who provide the guidelines needed to build high quality contents, as well as application designers and developers who are able to deliver the educational applications demanded by students. Thus, the success of applications is related to the organization of the team and assigned responsibilities [3].

Taking into account the need of developing educational applications with adequate user interfaces for multiple devices and software platforms, this work presents a cloud-based system that allows the generation of educational applications for smartphones, tablets, Web (Desktop Computers), and TV devices (Android-based TV Box and Smart TVs). It is noteworthy mention that, all user interfaces used by the application developed by the cloud-based platform complies with User Interface Design Patterns (UIDP). Several approaches are oriented to study UI patterns that can be applied in the development of desktop, Web and mobile applications. Our platform uses different mobile UIDP that have been identified in several studies [4, 5] for mobile application development. Also, several proposals on the use of digital interactive TV have been used for software development.

It is noteworthy mention that one of the most important aspects of our approach is the use of TV Android-Based Interface Design Patterns for software development. The cloud-based platform allows the development of user interfaces following the 10-foot UI specification that provides the criteria needed for the development of Digital-TV interface. This specification was adopted by Google as a set of design pattern to generate TV applications.

This paper is organized as follows: Sect. 2 describes a review of previous related works. Section 3 presents the platform architectural design. The case study that demonstrates platform functionality is included in Sect. 4. Finally, future work and conclusions are presented in Sect. 5.

2 State of the Art

In this section, we describe some related works about cloud computing. Cloud computing is a style of computing in which dynamically scalable and often virtualized resources are provided as a service over the Internet. Users do not need having knowledge of, expertise in, or control over the underlying infrastructure in the Cloud that supports the services rendered to them [6]. Due to proliferation of mobile devices such as smartphones and tablet computers, which have leaded to ubiquitous mobile applications, i.e., applications that can be accessed at any time and from anywhere over the Internet by using mobile devices, cloud computing has taken advantage of mobile computing. This is evidenced by the recent incursion of mobile operators such as Vodafone [7], Verizon and Orange on the cloud computing market. According to Fernando et al. [8], Mobile Cloud Computing (MCC) is still in its infancy; however, it could become the dominant model for mobile applications in the near future. March et al. [9] proposed a framework that merges mobile and cloud computing concepts for the development of rich mobile applications, a new generation of distributed mobile applications that provides rich functionalities.

Mishra et al. [10] proposed a mobile-based cloud-computing framework for the integration of mobile applications with cloud services aimed at processing and data storage. In the context of platforms that allows the generation of software applications, Srirama et al. [11] proposed a MCC middleware for mobile mash-up applications, i.e., applications that combines data from diverse cloud services. Dukhanov et al. [12] presents an approach for the design and implementation of Virtual Learning Laboratory (VLL) by using cloud computing technologies within the model of AaaS (Application as a Service).

Ercan [13] presented a survey to collect the required data for the use of cloud computing in the universities and other governmental or private institutions in the region of Turkey. This survey will help us to know the current status and probable considerations to adopt the cloud technology. Results are shown as a pie chart and the labels on each different slice represent different industrial sectors and services. Ozdamli and Bicen [14] present a quantitative method to determine the conditions that affect student's perception and competences, towards mobile learning using mainly cloud computing services. The author selected Dropbox because in Bicen and Ozdamli [14] study results showed that most of students prefer Dropbox cloud computing services the Education Faculty.

Saad and Selamat [15] presented an UPSI Learning Management System called MyGuru2. MyGuru2 is an e-learning portal that offers a robust set of teaching and learning tools, functions and features. The author proposed the cloud computing as a solution where MyGuru2 will be put on convenient, on demand network access to a shared pool of configurable computing resources that can be maintained and provisioned with minimal management effort. Lin et al. [16] proposed a cloud-based reflective learning environment to assist instructors and students in developing and strengthening reflection ability during and after actual class sessions. The authors presented an industrial course in a Taiwanese University and evaluated the proposal using several questionnaires, and interviews. Ivanova and Ivanov [17] identified the sequence of actions that the users perform for developing their applications and services in order to build a complete teaching/learning environment. Furthermore, in this study a model of the authoring process occurring in the cloud of Web 2.0 applications and services is developed. Sultan [18] presented several examples of cloud users to demonstrate the emerging popularity of cloud computing with some educational and business establishments. Finally, a case study of the University of Westminster was presented. Shakil et al. [19] proposed an effective framework for managing university data using a cloud based environment. The authors presented a new simulation framework that demonstrates the applicability of the cloud computing in the current education sector. The framework consists of a cloud developed for processing an universities database which consists of staff and students.

Kao et al. [20] proposed a Web-based runtime environment for cross-platform offline-able mobile and desktop Web applications. This environment called WOPRE relies on iGoogle™ which is an environment for displaying HTML5-JavaScript gadgets within a customized Google™ homepage where gadgets are created by using the Google™ Gadgets Application Programing Interface (API). Moreover, WROPE is composed of three main subsystems: (1) a subsystem responsible for offline content management and data synchronization that uses the Google™ Gears API, (2) a sub-system responsible for Web content adaptation allowing developers to explicitly select the sections of Web content that can be displayed in mobile versions of Web applications and (3) an application market that represents an interface between developers and end-users for publishing and installing applications.

3 Software Architecture Description

The proposed architecture is shown in Fig. 1. This architecture is composed of four layers; each layer fulfills a specific function by using a set of components whose tasks are described as follows:

Cloud Business Tier: This layer is addressed by the Web-based platform, i.e. teachers through a Web browser. Applications built using the tools in the lower layer are executed at this level. The entry point is the Request Handler component, which acts as a gateway receiving all the requests for user interactions.

Cloud Platform Tier: This layer is addressed by the cloud-based platform for software generation. It is designed for developing and deploying new applications through a wizard. In this sense, the main component of this layer is the Wizard component, which allows end-users to design, develop and publish (or deploy) mobile, Web, and TV applications. Furthermore, this layer provides programming language-specific wrapper components for cloud services APIs allocated at lower layer. According to the above, the *Cloud Platform Tier* implements a development

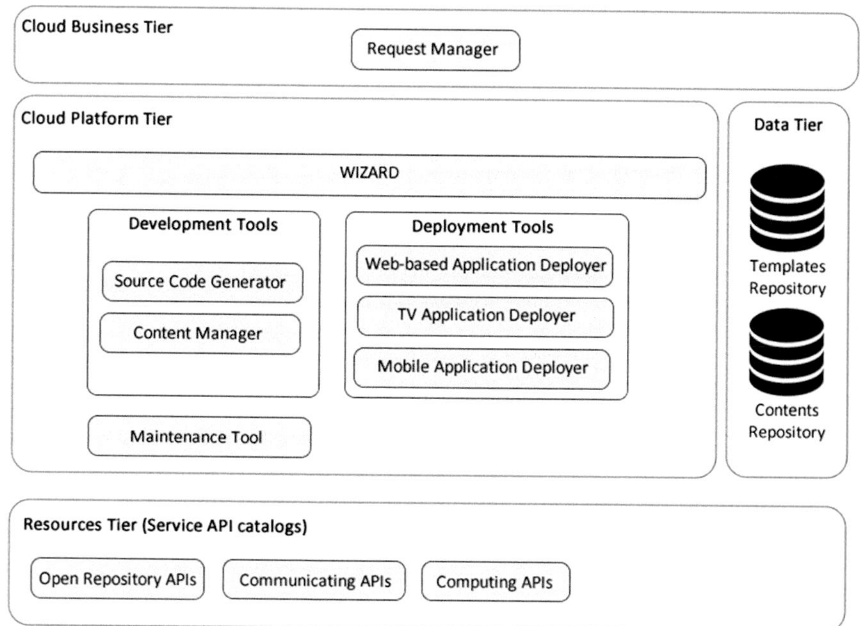

Fig. 1 Software architecture

process for multi-device applications which covers almost all the phases of the typical development life-cycle, namely, design, development, publishing (or deployment) and maintenance. Into this layer, there are set of components with a specific functionality that is explained as follow:

(1) *Source Code Generator*, which, depending on the required type of application is responsible for generating application directory structures, generating configuration files, and generating source code files. In fact, this component generates source code of rich mobile applications either in a deployable or installable for by using a set of templates.
(2) *Content Manager* is responsible to embed the desired educational contents in the applications developed.
(3) *Web-based Application Deployer* is responsible for deploying Web applications for desktop devices.
(4) *TV Application Deployer* is responsible for deploying Web applications for Smart TV and Android TV Box devices.
(5) *Mobile Application Deployer* is responsible for deploying Web applications for mobile devices such as tablets and smartphones
(6) *Maintenance Tool* is responsible for providing facilities to manage already deployed applications.

Data Tier: This layer is mainly aimed at storing XML-based documents by means of a set of repositories, which represents the core of the source code generation process. *Templates Repository* enables persistence for application front-end templates (by using UI Design Patters), and *Contents Repository* stores all educational contents used by applications. It is noteworthy mention that, the repository stores previously loaded contents under permissive licenses such as creative commons, this is in order to provide educational contents for teachers.

Resources Tier: This layer is mainly aimed to manage external repositories through the *Open Repository APIs* such as CK-12, Curriki, BBC, and Academic Earth to mention but a few. This task was done through the Communicating API and the Computing API.

With the aim of explaining the functionality of the open cloud-based educational platform, a case study of the generation of an educational application for TV is presented in the next section.

4 Case Study: Generation of an Educational Cloud-Based Course for Chemistry

Our cloud-based platform is a Web-based application that allows an author (teacher) to create an educational application for multiple devices such as tablets, smartphones, desktop computers (HTML5 Web-based applications), and TV (Android TX Box and Smart TV) by following a set of steps. By using this authoring tool, authors can build their own educational application to be displayed on all the previously mentioned devices, using only the contents that they need, or also, if the wanted, users can search for existing content within an open repository. The main functionalities of the cloud-based platform are to provide an author with a Web-based tool to be intuitive and easy to use to develop educational applications for multiple devices and platforms, which can reinforce academic subjects inside the classroom. In order to generate an educational application, the author must follow a set of steps. Therefore, a case study to understand the clod-based platform functionality is presented below. The case study describes the generation of a TV application for chemistry courses.

Let us suppose that a high school chemistry teacher wants to create a set of videos to support their lessons. In this context the teacher knows what type of information is required for a chemistry course. However, the information required is not available, so he needs to use Internet to search these contents. When the user finally accesses the contents, he would have a set of videos to be used separately, or he would need to carry out several activities to create his own set of videos. Under this scenario, there are some limitations, such as: (1) the teacher's lack of experience for searching and creating learning material from the Web, and (2) the teacher's unawareness of the standards for the development of TV applications. As solution to these issues, our cloud-based platform proposes (1) a set of educational contents stored in a repository that can be freely used, (2) a Web-based system to generate an educational application for TV following a set of steps and a set of guidelines for the development of TV interfaces. From this perspective, the teacher only needs to follow a set of steps described as follows:

(1) The teacher accesses by using an authentication mechanism.
(2) If the teacher is successfully authenticated, the platform will display the menu available on the main Web page. In this section the teacher selects the option to generate a TV application. Then, in this step wizard will be launched to generate a TV application.
(3) Once the wizard has been launched, the first step consists in selecting the content types to be included in the TV application. These content types can be audio, video, image, or text files. This step is depicted in Fig. 2.
(4) Once the content types to be included in the TV application were selected, the user must choose the main view or menu to be included in the application. To do this, the cloud-based platform includes different design menus for an application. In this step, the user has to select the menu design according to his preferences.
(5) Once the user has chosen the type of menu for the application, the next step is to add the content types. We have developed a set of templates to incorporate each one of the content types. Each template has been designed following the 10-foot UI design scheme. The order to select the content types is as follows: (1) if the user selected the four types of contents (audio, image, text, and video), the platform requests to introduce the audio contents. The user must select a template from the set of available templates for this content type; next he clicks on the "upload content" button. In this step, the user either chooses the location of an audio file on his computer, or looks up for content within the platform repository. This process is depicted in Fig. 3.

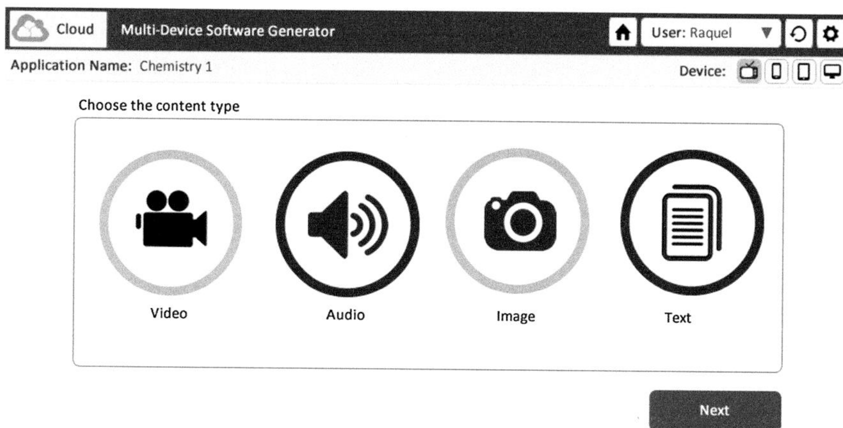

Fig. 2 Showing the selection of content types for TV application

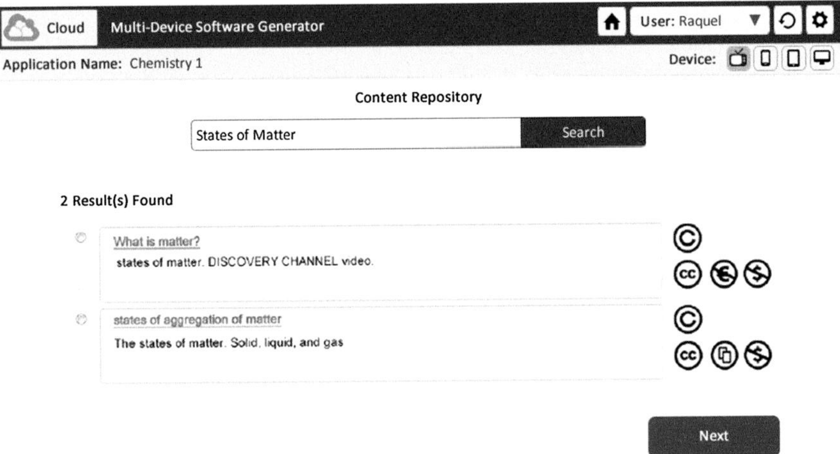

Fig. 3 Search and selection of content

If the user wants to upload other contents, he can continue selecting templates and uploading contents. Otherwise, the user can click on the "Ready" button to upload the following content type. This process is the same for uploading image, text, and video contents respectively. It is important to mention that some templates available allow the integration of more than one content type. When the user has finished adding the content types, he clicks on the button "Finish".

(6) The last step to generate the application consists in adding the links to each one of the templates chosen. To perform this, the platform presents a list with each one of the templates, and then the user must select the template to be linked. At the end of this step, the user must click on the "Finish" button so that the cloud-based platform starts compiling the application.

(7) While the application is being compiled, a progress bar is displayed to indicate the percentage of completion. Then, when the application has been successfully compiled, a confirmation dialog window is presented on the screen. This window asks the user if he wants to export the generated application. If the user clicks the "Yes" button, the file will be downloaded into a ZIP format.

(8) When the user has downloaded and decompressed the ZIP file, an .APK extension file is located, which contains the TV application. The next step will be therefore to install the application to be displayed on the TV. In this case study, the application is installed using a USB drive. Once the application is installed, the user can access each content through the menu. This process is depicted in Fig. 4.

This case study application was tested on the LG LCD TV 32″ (with Android TV-Box and the Samsung Smart TV, both having a HDMI connector. Other case studies have been proposed to generate TV-based applications for different academic topics and subjects such as bullying and discrimination, Mathematics, and Biology.

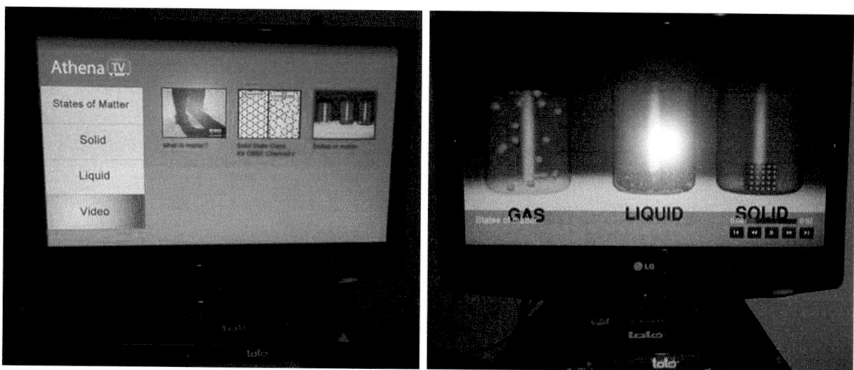

Fig. 4 TV application

5 Conclusions and Future Work

This work presented a cloud-based platform to generate educational applications for TV, Mobile, and Desktop applications by using User Interface Design Patterns. Our software architecture provides the necessary modules for the design of interfaces based on 10-foot UI design scheme and UI design patterns with the aims of bringing an appropriate user experience for the device used, and facilitating usability of the educational contents presented on Tablets, Smartphones, Computers, and TVs. It is important to mention that even though the cloud-base platform was designed to fill a need in an educational context, the system presented in this paper is not limited to educational applications. Any type of multimedia application to present content can be generated with the cloud-based platform by incorporating the appropriate content. In addition, we propose a scalable architecture design can be subject to new features such as content recommendation.

As future work, we are considering to scale the software architecture. New components and modules will be added for the development of TV educational applications using Interactive Digital TV technology combined with MHP standard, which supports several types of interactive applications. Moreover, since recommendation systems are being used in diverse contexts such as e-commerce, e-marketing, music, book, and movie recommendations, and news filtering among others, authors have also considered incorporating collaborative filtering techniques to generated software. This could reinforce students' acquired knowledge, or facilitate the comprehension of certain topics with the visualization of other contents that were useful to other users.

Acknowledgments This work was sponsored by the National Council of Science and Technology (CONACYT), the National Technology of Mexico (TecNM) and the Public Education Secretary (SEP) through PROMEP.

References

1. Dees, W., Shrubsole, P.: Web4CE: accessing web-based applications on consumer devices. In: Proceedings of the 16th International Conference on World Wide Web, pp. 1303–1304. ACM
2. Caswell, T., Henson, S., Jensen, M., Wiley, D.: Open content and open educational resources: enabling universal education. Int. Rev. Res. Open Distrib. Learn. **9** (2008)
3. Salinas, J.: La integración de las TIC en las instituciones de educación superior como proyectos de innovación educativa. En soporte digital, España (2002)
4. Neil, T.: Mobile Design Pattern Gallery: UI Patterns for Smartphone Apps. O'Reilly Media, Inc. (2014)
5. Tidwell, J.: Designing interfaces. O'Reilly Media, Inc. (2010)
6. Colombo-Mendoza, L.O., Alor-Hernández, G., Rodríguez-González, A., Valencia-García, R.: MobiCloUP!: a PaaS for cloud services-based mobile applications. Autom. Softw. Eng. **21**, 391–437 (2014)
7. http://www.vodafone.com/content/dam/vodafone/about/what/white_papers/connecting_tothe-cloud.pdf
8. Fernando, N., Loke, S.W., Rahayu, W.: Mobile cloud computing: a survey. Future Gener. Comput. Syst. **29**, 84–106 (2013)
9. March, V., Gu, Y., Leonardi, E., Goh, G., Kirchberg, M., Lee, B.S.: μcloud: towards a new paradigm of rich mobile applications. Proc. Comput. Sci. **5**, 618–624 (2011)
10. Mishra, J., Dash, S.K., Dash, S.: Mobile-cloud: a Framework of Cloud Computing for Mobile Application. Advances in Computer Science and Information Technology, pp. 347–356. Springer (2012)
11. Srirama, S.N., Paniagua, C., Flores, H.: Croudstag: social group formation with facial recognition and mobile cloud services. Proc. Computer Sci. **5**, 633–640 (2011)
12. Dukhanov, A., Karpova, M., Bochenina, K.: Design virtual learning labs for courses in computational science with use of cloud computing technologies. Proc. Comput. Sci. **29**, 2472–2482 (2014)
13. Ercan, T.: Effective use of cloud computing in educational institutions. Proc. Social Behav. Sci. **2**, 938–942 (2010)
14. Ozdamli, F., Bicen, H.: Effects of training on cloud computing services on M-learning perceptions and adequacies. Proc. Social Behav. Sci. **116**, 5115–5119 (2014)
15. Saad, M.N.M., Selamat, A.W.: UPSI Learning Management System (MyGuru2) in the cloud computing environment. Proc. Social Behav. Sci. **67**, 322–334 (2012)
16. Lin, Y.-T., Wen, M.-L., Jou, M., Wu, D.-W.: A cloud-based learning environment for developing student reflection abilities. Comput. Hum. Behav. **32**, 244–252 (2014)
17. Ivanova, M., Ivanov, G.: Cloud computing for authoring process automation. Proc. Social Behav. Sci. **2**, 3646–3651 (2010)
18. Sultan, N.: Cloud computing for education: a new dawn? Int. J. Inf. Manage. **30**, 109–116 (2010)
19. Shakil, K.A., Sethi, S., Alam, M.: An effective framework for managing university data using a cloud based environment. arXiv preprint arXiv:1501.07056 (2015)
20. Kao, Y.-W., Lin, C., Yang, K.-A., Yuan, S.-M.: A web-based, offline-able, and personalized runtime environment for executing applications on mobile devices. Comput. Stand. Interfaces **34**, 212–224 (2012)

Promoting e-Commerce Software Platforms Adoption as a Means to Overcome Domestic Crises: The Cases of Portugal and Spain Approached from a Focus-Group Perspective

Ramiro Gonçalves, José Martins, Jorge Pereira, Manuel Cota and Frederico Branco

Abstract Focus group interactions led to a set of strategic recommendations with regards to how to improve e-Commerce adoption levels in Iberian enterprises, essential in times of crises. Suggestions include: (1) Create actions to influence the governments of the Iberian Peninsula to re-evaluate the legislation that regulates e-Commerce; (2) Encourage venture capitalists, banks and business angels to create financing lines with better access; (3) Encourage European institutions of higher education to create partnerships with Iberian enterprises in a way that the technical know-how in these enterprises could be mixed with the scientific knowledge of those institutions; (4) Create, alongside with training organizations and universities, a set of new training courses directed at Iberian enterprises focusing on concepts such as Web 2.0 capabilities and the maintaining of a coherent online organizational identity. Increases in e-Commerce transactions may bring changes in mentality and greater prosperity to nations such as Portugal and Spain.

Keywords e-Commerce software platforms · Adoption · Focus group · Iberia

R. Gonçalves (✉) · J. Martins · J. Pereira · F. Branco
University of Trás-os-Montes e Alto Douro, Vila Real, Portugal
e-mail: ramiro@utad.pt

J. Martins
e-mail: jmartins@utad.pt

J. Pereira
e-mail: jorge.pereira@infosistema.com

F. Branco
e-mail: fbranco@utad.pt

R. Gonçalves · J. Martins · F. Branco
INESC TEC and UTAD, University of Porto, Porto, Portugal

M. Cota
University of Vigo, Vigo, Spain
e-mail: mpcota@uvigo.es

J. Mejia et al. (eds.), *Trends and Applications in Software Engineering*,
Advances in Intelligent Systems and Computing 405,
DOI 10.1007/978-3-319-26285-7_22

1 Introduction

The study of the Internet and its technologies as a means to do business worldwide has been in the agenda of several research projects [1]. The current era of Information and Communication Technologies (ICT) ushered in the era of electronic commerce and the information society; where traditional management paradigms are challenged and new business models are sought; and where the continuous improvement of both business and processes needs leads to a rise in the importance of their alignment [2, 3]. The adoption of new Internet-based business models will certainly generate positive business revenues, thus the potential for e-business is high enough to compel enterprises to go online [4–6].

The benefits in using e-Commerce as a new way for SMEs (about 99 % of the EU market [7]), to do business are well supported by the existent literature [8, 9]. Nevertheless, it's also very important to understand the variables inherent to the adoption of e-Commerce websites, in order to better adjust a company's online presence and raise its business volume [10–12].

The adoption of e-Commerce by organizations is a very up-to-date issue. Despite the existence of innumerous valid empirical works [13–15], to our knowledge there are few others performing qualitative analysis to the e-Commerce adoption issue, thus similar to the one presented in this article, and revealing indicators on the adoption of e-Commerce by organizations all around the World [16–18]. As concerns Iberia (the peninsula including Portugal and Spain) we can see two different indicators regarding e-Commerce adoption with distinct results, which are the number of Iberian organizations using e-Commerce and the number of Iberian Internet users. These indicators both present themselves with relatively low values and, the reasons behind those factors, are still to be explained. With this in mind, we used a focus group in order to achieve a global set of recommendations (and expected results of their implementation) that could be used for improving the adoption of e-Commerce in Portugal and Spain. Besides reaching the referred set of recommendations, we have also created a strategic proposal to encourage Iberian e-Commerce adoption. This proposal is composed by several action guidelines that aim to offer some guidance on how to solve the referred adoption issue.

2 Research Characterization and Conceptualization

2.1 *Research Objectives*

Although the theme of e-Commerce has already been a subject of discussion by various authors over the last 15 years, the problems inherent to its adoption still need to be overcome. This situation is proven by several reports published by the EU, which show that the number of active e-Commerce initiatives is low and that the level of adoption to those same initiatives is far from the desired levels.

Although the values at the European level are not as good as expected, with regards to Iberian e-Commerce initiatives, the values are even somewhat lower.

2.2 Research Question

The research question addressed was as follows: What paths exist to increase the adoption to e-commerce transactions, in Portugal and Spain, as a means to contribute to the overcoming of the domestic crises in these countries?

2.3 Adoption of Electronic Commerce in Portugal and Spain

Portugal is characterized as a society where profound change has occurred, especially over the last two decades, which has seen "the emergence of new values" [19]. Has "the aging of the population with an increase in the number of people aged over 65 years [with] a fall in birth and fertility rates" had a profound effect on e-Commerce adoption levels in Portugal? Has the "massive transfer to the tertiary sector", away from the primary sector and from the industrial sector, led people away from e-Commerce? Are the low figures for Portuguese women dedicated to technical-scientific and administrative professions, the lowest registered in the EU, at 11 and 15.2 %, respectively, of the active female workforce, testimony to a less-developed society which shies away from e-Commerce?

Conversely, the Spanish Kingdom, centered around Madrid, "has approached and overcome important challenges, risks, and changes in a very deep and widely spread democratization process" [20]. Spain is no longer isolated, from a dictatorship it has moved to being currently rather open, tolerant and decentralized, with European-minded people for the most part, though nationalist views still prevail in some quarters [20]. Deeply held Spanish beliefs include, however, "a bureaucratic approach: by observing as many norms and procedures as might be available because this is a strategic approach to avoid risks and uncertainties" [20]. Could this be a reason behind lagging e-Commerce figures in Spain, a desire for certain transactions without risk?

Several studies have been recently done, in both Portugal and Spain, regarding the use of the Internet for e-Commerce and e-business purposes [21, 22]. The results obtained claimed that there is a lack of comprehension of the motives for low levels of adoption to e-Commerce in these two countries, though we recognize that some indicators show a somewhat positive trend.

A total of 16 % of Portuguese companies have adopted the Web as a new commercial channel, against 10 % in Spain and 12 % in the EU. For these companies in Portugal this channel represents 12 % of total income, against 10 % in

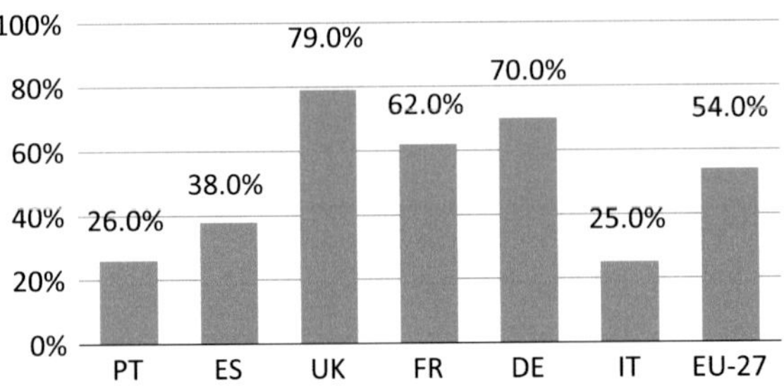

Fig. 1 Percentage of Internet users that have ordered goods or services over the Internet for private use during 2011 (Adapted from: [23])

Spain and 13 % in the EU—which is in fact difficult to understand, regarding the percentage of companies selling online. Also, the percentage of Portuguese organizations buying on the Web is 19 % against 18 % in Spain and 24 % in the EU. The Portuguese have a low rate of Internet shopping adoption, with only 26 % of Internet users having already made some purchases on the Web, against 54 % in the EU [21].

According to EU [21], 55 % of all Portuguese enterprises use applications for integrating internal business processes and in Spain this figure is at 52 %, which is more than in the EU (41 %). 32 % of Portuguese companies automatically exchange business documents with customers/suppliers and in Spain this figure is at 12 % (the EU average is at 26 %). Regarding information sharing, 31 % of Portuguese companies share information electronically with customers/suppliers in Supply Chain Management and in Spain this figure is at 14 % (the EU average is at 15 %). By including in this analysis Eurostat's "Internet use in households and by individuals in 2011" study, one can observe that the number of Internet users ordering goods or services has grown from 47 to 54 % in the period between 2006 and 2011, while during the same period this indicator only reached 26 % in Portugal and 38 % in Spain (Fig. 1).

In Fig. 1 we can see that, as concerns business-to-consumer (B2C) purchasing rates over the Internet, Southern European countries such as Portugal (PT), Spain (ES) and Italy (IT) are well below the average for the EU-27. The UK, France (FR) and Germany (DE) show a more positive trend.

3 Data Collection and Analysis

Aiming to answer the research question mentioned above and in order to envision a way forward for e-Commerce, in Portugal and Spain, we decided to start a focus group process, in which a number of specialists were involved.

3.1 Research Methodology

In order to gain further insights relating to the e-Commerce activities of Iberian enterprises, and namely into how adoption levels to e-Commerce could be improved, we conducted a focus group research project, a qualitative research method popular for being able to provide detail [24, 25].

Focus group research involves gathering a group of carefully selected individuals (i.e. not randomly selected) to discuss various topics of interest [26]. On the present research eight specialist were selected (following a Curriculum Vitae analysis which we performed), the group thus having an appropriate size within what according to Ghauri and Gronhaug [24] and Gonçalves et al. [27] is deemed fitting—namely between six to ten members.

The main goal in using a focus-group approach was to use its elements to help us interpret and reap practical results concerning the consequences and implications of what we had found in our studies to date [24]. The research project plan followed can be seen in Fig. 2.

The focus group members did not receive any payment for their collaboration, which occurred in an environment of considerable e-Commerce knowledge sharing, during three separate focus group meetings, leading up to the final discussion and creation of the e-Commerce adoption strategy proposal.

3.2 Focus Group Results and Considerations

After the initial discussion on the considerations that would lead towards an improvement in e-Commerce adhesion, the focus group discussed what the positive or negative arguments were for each of these same considerations. In Table 1 we present the list of variables that, in the focus group experts' opinions, will have a decisive impact on the adoption of e-Commerce software platforms. These variables are organized according to their scope (social, business or technology) in order to allow for a better understanding of each of the variables.

After classifying each variable according to their positive or negative impact on e-Commerce software platforms adoption, the focus group members also assumed that a set of global considerations should be reached in order to provide for an improved understanding on what the variables represent and what might be their overall impact.

These considerations were also grouped according to their scope: Social considerations; Business Considerations; Technological Considerations.

After achieving Table 1 set of variables, we requested the focus group to describe their point of view in an informal manner on each of the variables, hence

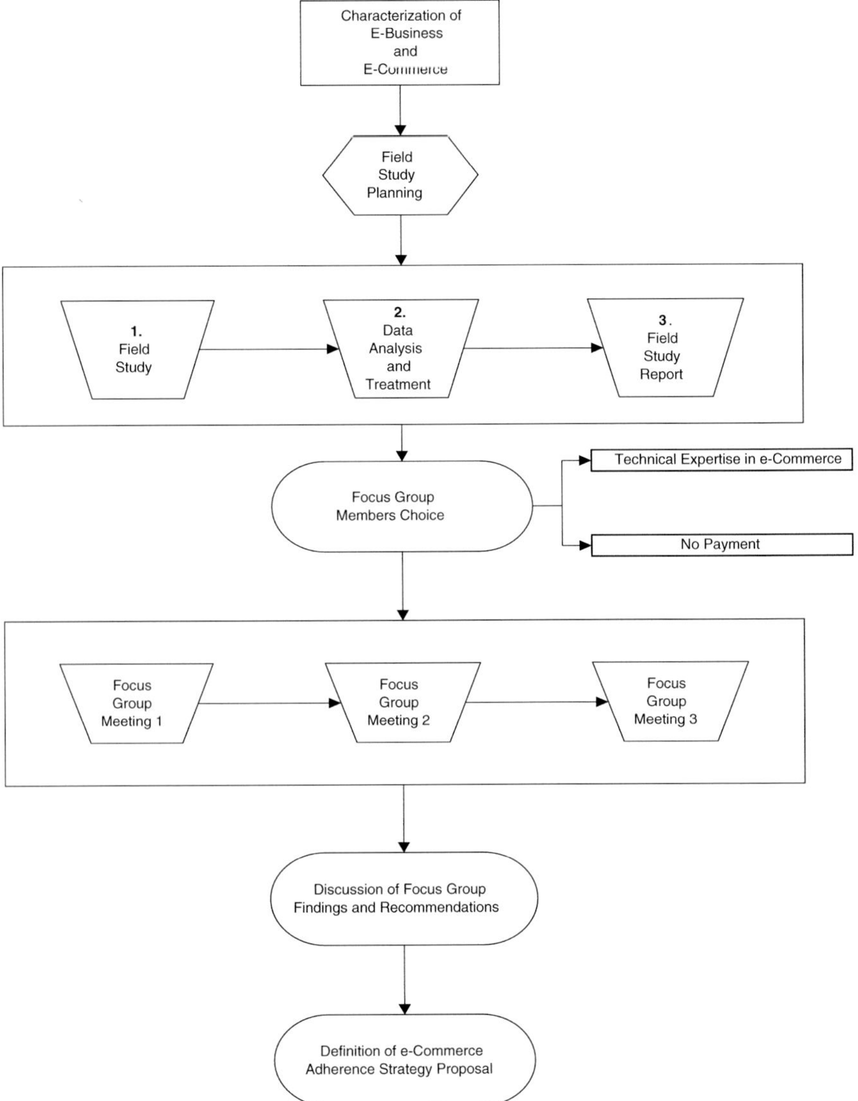

Fig. 2 Research project schema

allowing to achieve an overall perspective that permitted to create a strategic proposal towards the adoption of e-commerce that is presented in the following chapter.

Table 1 List of variables that may impact e-Commerce software platforms adoption

Social	Business	Technology
Privacy	ICT training	Security
Reputation	Expectations	Usability
e-Commerce knowledge	Experience with logistics	Maturity
Cultural issues	Support for creating e-Commerce initiatives	Expensive and inadequate solutions
Fear of use	Price	Available software packages
Information available		Software requirements

4 Strategic Proposal Aiming to Increase e-Commerce Software Platforms Adoption in Iberia

As was the aim, from the beginning of the present project, after analyzing the interactions made by the focus-group and the recommendations that resulted from it, a new set of ideas and possible solutions for the specific issues was reached. By discussing and re-arranging the referred solutions and ideas, we managed to create a strategic proposal for e-commerce software platforms adoption composed by eight guidelines that were especially designed to meet the entire spectrum of the previously identified problems. Our strategic proposal incorporates a schematic representation of seven key points that, in our opinion, when correctly related, can provide a positive contribution to e-commerce software platforms adoption (Fig. 3).

Improve the Overall Communication Between all Actors: The World of business is composed by several distinct actors, according to what their motivation and skills are, and who can each contribute, at every step, to the development and growth of this World. The actors who can be more easily recognized in this context are certainly companies and their suppliers, customers, legislators and regulators, as well as scientific institutions (such as universities) that leverage, in part, the necessary scientific know-how.

Promote the Establishment of Business Partnerships: When an Iberian organization intends to carry out an e-Commerce initiative or to improve an already existent one, it should take advantage of the referred to multi-disciplinary market and create ventures in which the partners' knowledge is shared. This sharing of knowledge, products and services will bring, of course, changes to its business processes and to the products/services it provides, through its e-Commerce initiatives, thus leading to improvements in the levels of adoption of this type of "business trade".

Increase the Number of Credible and Valid Investment Lines: The number of Iberian Small and Medium Enterprises (SME) is far superior to the number of large enterprises. This said, it is possible to perceive that the majority of Iberian

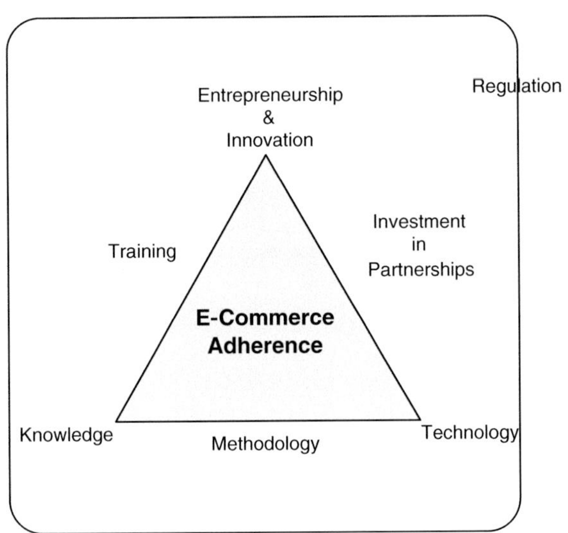

Fig. 3 Strategic proposal key points schema

enterprises present several limitations, including ones related to the availability of capital. With the referred limitations in mind, in order to create or maintain good quality e-Commerce initiatives (capable of attracting and maintaining customers), Iberian enterprises have a serious need for investment.

Encourage Changes to Regulation Mechanisms: If the existent regulation mechanisms could be changed in order to become more in sync with the global reality of e-Commerce, it would be easier for enterprises all across the globe to understand what regulations they had to fulfill and how to do so.

Promote Methodologies for Implementing e-Commerce Initiatives That Have Already Provided Good Results: It is very important that Iberian companies make use of proven methodologies when combining internal organizational knowledge with their technology know-how in order to maximize their chances of success [13, 28]. Benchmarking processes should be included so as to constantly identify internal and external best practices appropriate to e-Commerce.

Encourage the Use of Personalized Functionality on e-Commerce Web Sites That Promotes the Development of a Long Term Electronic Relationship With Iberian Peninsula Users: High uncertainty avoidance and the importance of interpersonal relationships [29, 30], which are cultural elements, discussed in the focus group interactions, need to be focused on to increase the adoption in Portugal and Spain to e-Commerce. Thus, e-Commerce web sites in Portugal and Spain may need to expertly and explicitly promote a feeling of preferential treatment amongst users from these countries.

Promote the Development of an Organization's Internal Knowledge: The complexity of an organization's business processes requires a high level of technical and management knowledge that must, at all times, be maintained. By enforcing the development and improvement of an organization's internal knowledge, managers and collaborators create mechanisms to facilitate the emergence of innovative initiatives and the improvement of internal technical skills, that when combined may lead to an increase in e-commerce software platforms adoption levels.

Induce the Provision of Training in e-Commerce to Executives and Professionals: Both Portuguese and Spanish organizations have in their staffs a great disparity as concerns technical skills. In order for a given organization to be successful, its collaborators must be as well qualified as possible. This will bring advantages such as the ability to create better products and supply better services, but also the ability to devise innovations and drive entrepreneurship business initiatives. It is important that Iberian organizations focus on supplying their staffs with the required training in order for them to carry out their work in a more efficient and successful manner. Quite specifically, adopting Web 2.0 related technologies and resources, such as social networks sites, will lead to greater empathy with customers and thus greater economic returns [31, 32].

Encourage an Entrepreneurial and Innovative Spirit in the Development of Technological Solutions: Innovation and entrepreneurship are decisive factors for the success of e-Commerce organizations given that they allow the adaptation of old business processes to the existent reality and also the deployment of creative and promising e-Commerce initiatives that may provide business improvements to organizations.

5 Limitations and Future Work

The use of a Delphi study as the technique to indulge the chosen qualitative methodological approach is yet to exist without faults and omissions, mainly supported with the argument of one group's opinion might not the same as other groups in a similar context. With this in mind, in a future work we would like to perform a Delphi study in which a broader and wider set of experts could be invited to participate and give their consensual opinion on the results achieved with the present research.

One other issue concerning this research is the inexistence of a wider validation to the achieved results, particularly by those who are directly vised in its content, the firms. Drawing on this limitation, and assuming Venkatesh et al. [33] and Martins [34] argument that the best methodological approach to follow when studying the IT/IS area of knowledge is the mixed methodology, in a future work we would like to performed an empirical study based on a survey addressed to organizations and aiming on validating the importance and relevance of the identified recommendations.

6 Conclusions

After a review of some of the existent literature and aiming to increase the success of e-Commerce investments, we conducted a Focus Group research project with Portuguese and Spanish specialists in various areas related to e-Commerce. This process began by analyzing the current e-Commerce situation and resulted in a set of recommendations that, if implemented, could contribute to improve not only the Iberian e-Commerce initiatives adoption levels but also their return on investment. Nevertheless, the great diversity of options and multiple cases of success (and failure) suggest that the aforementioned initiative success may only happen if a proper combination of all the presented recommendations is made.

As a way to implement the presented recommendations and to increase Iberian e-Commerce initiative adoption levels, we have created a strategic proposal composed of nine guidelines that were designed to show a possible path aiming towards a not-so-distant future with regards to how to carry out good quality and highly successful e-Commerce initiatives.

References

1. Ifinedo, P.: An empirical analysis of factors influencing internet/e-business technologies adoption by SMEs in Canada. Int. J. Inf. Technol. Decis. Making **10**(4), 731–766 (2011)
2. Trkman, P.: The critical success factors of business process management. Int. J. Inf. Manage. **30**, 125–134 (2010)
3. Pereira, J., Martins, J., Gonçalves, R., Santos, V.: CRUDi framework proposal: financial industry application. Behav. Inf. Technol. **33**(10), 1093–1110 (2014)
4. Morais, E., Pires, A., Gonçalves, R.: e-Business maturity: constraints associated with their evolution. J. Organ. Comput. Electron. Commer. (2011)
5. Branco, F., Gonçalves, R., Martins, J., Cota, M.: Decision support system for the agri-food sector—the Sousacamp group case. In: World Conference on Information Systems and Technologies. AISTI, Azores, Portugal (2015)
6. Martins, J., Gonçalves, R., Pereira, J., Cota, M.: Iberia 2.0: a way to leverage Web 2.0 in organizations. In: 7th Iberian Conference on Information Systems and Technologies (CISTI 2012), pp. 1–7, Madrid, Spain (2012)
7. EU: small and medium-sized enterprises (SMEs)—fact and figures about the EU's small and medium enterprise (SME). Policy highlights overview. EU (2011)
8. Grandon, E.E., Nasco, S.A., Mykytyn Jr, P.P.: Comparing theories to explain e-Commerce adoption. J. Bus. Res. **64**, 292–298 (2011)
9. Solaymani, S., Sohaili, K., Yazdinejad, E.: Adoption and use of e-commerce in SMEs. Electron. Commer. Res. **12**, 249–263 (2012)
10. Gonçalves, R., Barroso, J., Varajão, J., Bulas-Cruz, J.: A model of electronic commerce initiatives in portuguese organizations. Interciencia **33**, 120–128 (2008)
11. Thomas, B., Simmons, G.: e-Commerce adoption and small business in the global marketplace: tools for optimization. IGI Global, Hershey (2010)
12. Goncalves, R., Branco, F., Martins, J., Santos, V., Pereira, J.: Customer feedback and Internet: means used by the biggest Portuguese companies. In: 2011 Proceedings of the International Conference on e-Business (ICE-B), pp. 1–4 (2011)

13. Chatterjee, D., Grewal, R., Sambamurthy, V.: Shaping up for e-Commerce: institutional enablers of the organizational assimilation of web technologies. MIS Q. **26**, 65–89 (2002)
14. Hadaya, P.: Determinants of the future level of use of electronic marketplaces: the case of Canadian firms. Electron. Commer. Res. **6**, 173–185 (2006)
15. Oliveira, T., Dhillon, G.: From adoption to routinization of B2B e-Commerce: understanding patterns across Europe. J. Glob. Inf. Manag. **23**, 24–43 (2015)
16. Alam, S.: An empirical study of factors affecting electronic commerce adoption among SMEs in Malaysia. J. Bus. Econ. Manage. **12**, 375 (2011)
17. Grandón, E.E., Nasco, S.A., Mykytyn Jr, P.P.: Comparing theories to explain e-Commerce adoption. J. Bus. Res. **64**, 292–298 (2011)
18. Simmons, G., Armstrong, G., Durkin, M.: An exploration of small business website optimization: enablers, influencers and an assessment approach. Int. Small Bus. J. **29**, 534–561 (2011)
19. Jesuíno, J.: Leadership and culture in Portugal. In: Chhokar, J., Brodbeck, F., House, R. (eds.) Culture and Leadership Across The World—The GLOBE Book of in-depth Studies of 25 Societies, pp. 583–621. Psychology Press, Routledge (2008)
20. O'Connell, J., Prieto, J., Gutierrez, C.: Managerial culture and leadership in Spain. In: Chhokar, J., Brodbeck, F., House, R. (eds.) Culture and Leadership Across the World—The GLOBE Book of in-depth Studies f 25 Societies, pp. 623–654. Psycology Press, Routledge (2008)
21. EU: Europe's digital competitiveness report. European Union (2010)
22. Morais, E., Gonçalves, R., Pires, J.: Electronic commerce maturity: a review of the principal models. In: IADIS International Conference on e-Society, Lisbon, Portugal (2007)
23. Eurostat: Eurostat community survey on ICT usage by households and by individuals. EU (2011)
24. Ghauri, P., Gronhaug, K.: Research Methods in Business Studies, 4th Edn. Prentice Hall, New York (2010)
25. Bernard, H.R.: Research Methods in Anthropology: Qualitative and Quantitative Approaches. AltaMira Press, Thousand Oaks (2005)
26. Kotler, P., Keller, K.: Marketing Management, 13th Edn. Prentice Hall, Englewood Cliffs (2008)
27. Gonçalves, R., Martins, J., Pereira, J., Oliveira, M., Ferreira, J.: Accessibility levels of Portuguese enterprise websites: equal opportunities for all? Behav. Inf. Technol. **31**, 659–677 (2011)
28. Epstein, M.J.: Implementing successful e-Commerce initiatives. Strateg. Financ. **86**, 23–29 (2005)
29. Takac, C., Hinz, O., Spann, M.: The social embeddedbess of decision making: opportunities and challenges. Electron. Markets Int. J. Netw. Bus. **21**, 185–195 (2011)
30. Hofstede, G.: Culture's consequences: comparing values, behaviours, intitutions, and organizations across nations. SAGE Publications Inc., Thousand Oaks (2001)
31. Fonseca, B., Morgado, L., Paredes, H., Martins, P., Gonçalves, R., Neves, P., Nunes, R., Lima, J., Varajão, J., Pereira, A., Sanders, R., Barracho, V., Lapajne, U., Rus, M., Rahe, M., Mostert, A., Klein, T., Bojovic, V., Bošnjak, S., Bošnjak, Z., Carvalho, J., Duarte, I., Casaramona, A., Soraci, A.: PLAYER-a European project and a game to foster entrepreneurship education for young people. J. UCS **18**, 86–105 (2012)
32. Martins, J., Gonçalves, R., Oliveira, T., Pereira, J., Cota, M.: Social networks sites adoption at firm level: a literature review. In: CISTI'2014—Iberian Conference on Information Systems and Technologies, Barcelona, Espanha (2014)
33. Venkatesh, V., Brown, S., Bala, H.: Bridging the qualitative-quantitative divide: guidelines for conducting mixed methods research in information systems. MIS Q. **37**, 21 (2013)
34. Martins, J.: Adoção de Redes Sociais Online - Um estudo sobre os fatores que afetam a sua adoção ao nível das empresas. Departamento de Engenharias, vol. PhD. Universidade de Trás-os-Montes e Alto Douro (2014)

Multi-tabletop System to Support Collaborative Design Assessment

Gonzalo Luzardo, Vanessa Echeverría, Yadira Quiñonez and Roger Granda

Abstract This paper presents a study that describes the design and implementation of a tabletop system for supporting collaborative design in the classroom. A case study and two experiments are presented in order to evaluate the usefulness of the proposed system for students and educators. Ten educators and fifteen students participated in the experiments. Findings show that the usefulness, as well as the easiness of the proposed system are perceived as good from both, students and educators. These results suggest that the proposed system does have potential to be used in other educational areas or as a baseline for similar approaches.

Keywords Collaborative design · Collaborative assessment · Multi-tabletop system

1 Introduction

Developing collaborative working skills is an important aspect for the academic training of students. Employers consider these skills as a fundamental requirement when hiring professionals [1, 2]. Several research have been directed at developing

G. Luzardo (✉)
Facultad de Ingeniería En Electricidad Y Computación,
Escuela Superior Politécnica Del Litoral, Guayaquil, Ecuaduor
e-mail: gluzardo@espol.edu.ec

V. Echeverría · R. Granda
Centro de Tecnologías de Información, Escuela Superior Politécnica Del Litoral,
Guayaquil, Ecuaduor
e-mail: vecheverria@cti.espol.edu.ec

R. Granda
e-mail: roger.granda@cti.espol.edu.ec

Y. Quiñonez
Facultad de Informática Mazatlán, Universidad Autónoma de Sinaloa, Sinaloa, Mexico
e-mail: yadiraqui@uas.edu.mx

J. Mejia et al. (eds.), *Trends and Applications in Software Engineering*,
Advances in Intelligent Systems and Computing 405,
DOI 10.1007/978-3-319-26285-7_23

proposals for computer-assisted technology, to promote and improve participation and collaborative student learning [3–5]. However, the way that people work in this and other collaborative environments has not changed significantly over the years. Some studies show that pen and paper still remains widely used in designing [6]. Technology has failed to displace traditional tools mainly for the following reasons: space flexibility; easiness of communication between individuals; and portability [6]. While collaborative work in the traditional way (e.g. on a board or paper) facilitates face to face communication between individuals, it also leads to some drawbacks such as difficulty in documenting the final work, replicate or share these works in digital repositories (it is often desirable to observe a group work) [7].

In a classroom, the traditional tools also create certain limitations; for example, difficulty to monitor the process of preparation and subsequent assessment of collaborative work, as educators usually only have the final version of these. This is a problem because educators might find it difficult to assign a rating, as well as meet the workload invested by students, their individual contributions and the quality of these [4]. Researchers have found that the perception of an unfair assessment should be taken into account in issuing a subject, since it is predictive of motivation, learning, and even aggressiveness that students show towards a particular subject [8].

In recent years, the scientific community has developed some research progress in the implementation of new technological tools, which are intended to facilitate collaborative tasks and solve partially the aforementioned drawbacks. Some authors have developed project of multi-touch surfaces into classroom [5, 9], to facilitate and support collaborative learning among a group of students, in order to lead their own learning, driving engagement, participation and creativity with the use of this technology. In another work, Mercier et al. [10] have developed an interactive board to run experiments with students to support both fluency and flexibility in mathematics called NumberNet; similarly, in [11] they proposed a visual application in multi-touch tabletop for preschool Mathematics called MEL-Vis. Both works present a workable solutions through qualitative and quantitative analysis.

Martinez et al. [12] present a solution called "Collaid". This work has been developed in order to enhance learning and teamwork. Collaid uses a touch screen and support for the participation of individuals in the design of conceptual maps. In addition, it uses information about the position of each person and their verbal interaction, in order to help determine the contribution of each individual and display a list of suggestions on topics that each user might want to use. These users feed information to their collaborative work in Collaid, using their fingers to type on a virtual keyboard. This solution is integrated with a monitoring component of collaborative work, which is used by the orchestrator of the work (the educator) to know the state of development of the work at all times.

Although there are several proposed solutions whose aim is to assist the collaborative work, some problems must be solved yet. Issues such as restricting the size of the touch surface using previous solutions, traceability of tasks, the ability to monitor the development of the task, the number of people who can participate, configuration complexity of the work environment, the cost of implementation, and tools that emphasize design software, do necessary research in this field.

Currently, in the literature there are four classes of digital tabletop systems: digital desks [13], workbenches [14], drafting tables [15], and collaboration table [16]. This work focuses on the design of collaboration tables, it describes the design and implementation of a low cost multi-tabletop system (MTS) using computer vision, to improve the analysis of work in a student's group. In addition, a case study is presented in order to validate the proposed system. This paper is structured as follows: Sect. 2 briefly describes the system structure. Section 3 presents the proposed implementation. Section 4 describes two experiments used to test the system and the obtained results. Finally, conclusions remarks at Sect. 5.

2 Multi-tabletop System Proposed

2.1 *Functional Requirements*

The following characteristics were identified for the proposed system with some of them, taken from other similar systems and include:

- *Multi-user support*: It allows multiple users to use the system simultaneously [17, 18].
- *Use of tangibles*: It provides digital and physical objects to the users that allows a more natural interaction with the system [19].
- *Freedom of movement move and work-space regulation*: It allows users free movement when they use the system. This avoids limiting the creativity of users when static workplaces are defined [20].
- *Interconnection with devices*: It allows data entry through mobile devices like cell phones or tablets [17].
- *Color based contribution distinction*: It allows a person to identify users' contributions by assigning them a color per user.
- *Monitoring and storage user actions*: It provides tools for monitoring and recording activities performed by the users working in multiple groups at the same time [21].

2.2 *Physical Design*

Figure 1 depicts an upper view of the physical design of the proposed MTS. A front projection design was chosen. Both, infrared camera and projector are attached to a tripod. The infrared camera aims to detect the position of the pens on the table (interaction area). Students can use pens or mobile devices to interact with the elements of the collaborative application projected on the table. Each action is tagged by a unique color per student. Students can also use tablets to input data to the MTS. A server is used to process data from infrared camera. Computer vision algorithms are used to identify and tag the pen's movements on the table. All data

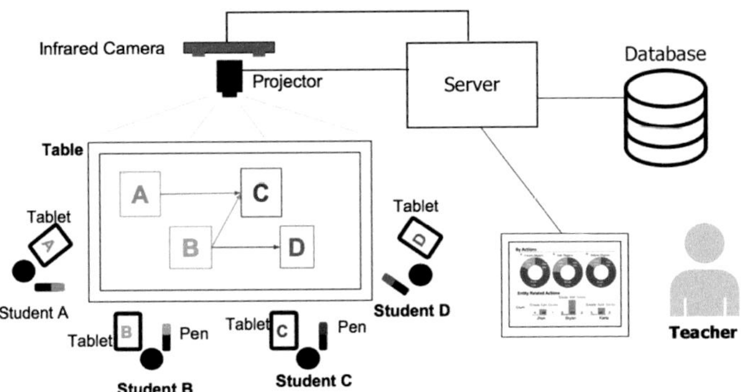

Fig. 1 Physical design of the student side of the proposed MTS

(objects, its history, student's interaction) are stored in a database. The communication between the tablets and servers is through a private wireless network. On the educator's side, the system allows monitoring student progress in collaborative tasks through a Web application running in the server.

The use of the computer vision approach, in contrast to others such as pressure sensors or capacitive surfaces, results in a portable and lower cost solution.

2.3 System Components

Four main components are identified for the proposed MTS:

- *Pen tracking*: It uses information provided by the infrared camera to detect and calculate their position and orientation on the table. Each pen has three infrared markers situated in different locations one from another in order to distinguish them. The information about the position of each pen is sent to the Visualization component and collaborative control through a TCP/IP protocol.
- *Mobile data entry*: It is responsible for creating and displaying a graphical user interface for the tablets, which allows data entry. It is used by the collaborative application so that the user can enter text or control application components. It is able to differentiate data entry made by each student through an authentication system. This information is sent to the Visualization component and collaborative control through a TCP/IP protocol.
- *Visualization and collaborative control:* It uses the information provided by the previous components to interact with the collaborative application. This contains control elements and objects that could be manipulated by the users using the pens or mobile devices. Pens movements received from the Pen tracking component are translated to mouse events. Plain text entry and component selection are received from the Mobile data entry component. Additionally, this component has a gesture recognition module to facilitate the user's data input.

All information about the state of the work made by students is stored in a file database. The database contains data about the users (identification number, tag color) and actions (create, edit or delete objects or components in the collaborative) performed by them stored in sequential order using the following format:

```
User     {user id, tag color in RGB format}
Actions  {user id, object type, object attributes, actions
         performed, time stamp}
```

- *Monitoring*: Uses the information stored in the file database to create useful text and graphical information about the collaborative task made by students using the MTS. The time stamp information is used to make an animation of how the objects and components of the collaborative application were created and manipulated over time by users.

3 Implementation

The proposed MTS is a combination of hardware and software, including a Web-based application for the educator's view. Students interact with the collaborative application using pens and tablets. Figure 2 shows an upper view of the

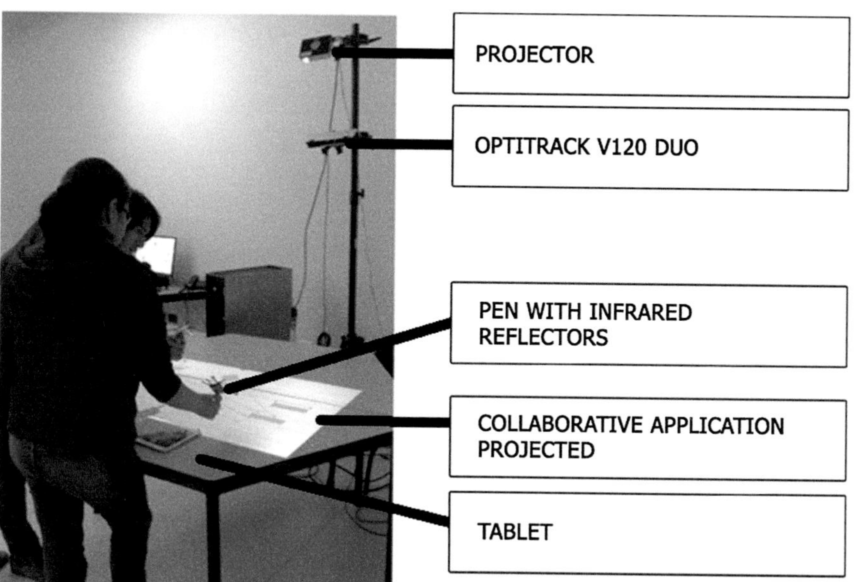

Fig. 2 Implementation of the student side of the proposed MTS

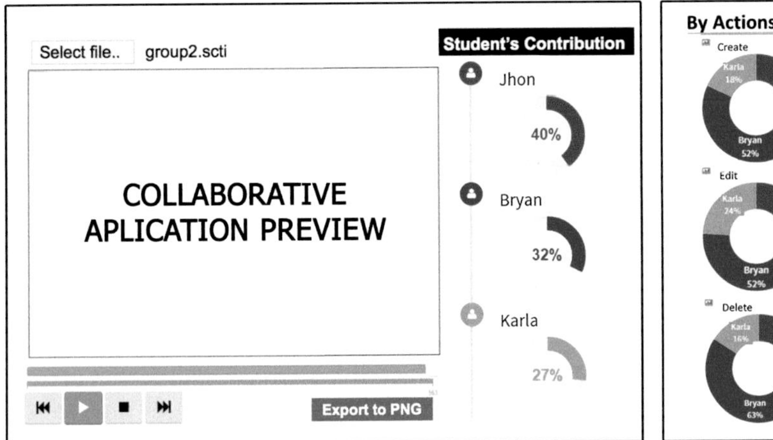

Fig. 3 Tools provided to educators for monitoring student's progress

physical scheme of the student side of the system. On the other hand, the educator only needs a device with a browser to access the system. Figure 3 shows some tools provided to educators in order to monitoring student's progress in collaborative tasks.

3.1 Hardware Implementation

The MTS was implemented using the following hardware components: a portable projector camera system that works with an Optitrack Motion Tracking V.120 Duo System and a mini projector (Aaxa Technologies P300 Pico projector). Other hardware components of the system are: a computer with CoreI5 2.9 GHz processor —4 GB RAM—500 GB HDD, a Samsung Galaxy Tab 3, and a pen with three infrared reflector markers.

3.2 Software Implementation

Tracking component: The tracking component keeps track of pens' positions, which are provided by the Optitrack's library Camera SDK. When students draw on the table, a touch event is generated through the TUIO (Tangible user interface) protocol [22], and then sent to the Visualization component and collaborative control.

Mobile data entry component: This was implemented as a responsive Web application, it allows users to authenticate to the system and then interact with collaborative application using a mobile device as a cell phone or tablet. This was developed in Python using Django Framework.

Visualization component and collaborative control: This component was implemented on the open source framework Multitouch for Java (MT4J) [23]. For gesture recognition, each stroke made with pen is processed by the PaleoSketch library [24]. The communication with collaborative application was implemented using multitouch events through the TUIO protocol. Every action on the tabletop and its related information is stored in a JSON file (file database). Elements entered by students are stored using a different colour to identify each member of the group.

Monitoring component: This component was implemented as a Web application using Python and Django framework. This component parses the information stored in the JSON file and show relevant text and graphical information about students' activities performed during the collaborative task.

4 Case Study: Database Model Collaborative Design

Database design is a topic covered in the Database Systems I, a course at a Computer Science Program. Regular activities in this course correspond to the design of a database system through real-life case studies, using an Entity-Relationship diagram as a part of the design. Being more specific, design activities are performed in groups from three, up to five students. Thus, it is common that the educator should keep tracking the design process from 4 to 8 groups simultaneously while doing the collaborative activity. As can be perceived, it is difficult for the educator to give a personalized and well established feedback once the activity has finished. In addition, it is common to see that work group students don't equally participate in the collaborative activity, even though they had the same grading points as their co-workers.

4.1 Experimental Setup

Two usability experiments were designed to test the usefulness of the MTS. Fifteen students enrolled in Database Systems I course (second academic semester—2014), were invited to participate in this experiment, as well as 10 educators (4 female, 6 male) with knowledge in topics covered in data modelling and 10 years of teaching experience.

Experiment 1: Ten educators participated in a pre-posttest experiment. Educators were asked to fill in a questionnaire, which contains three questions to measure the following variables: ease of grading individuals and groups; and, equality of individual's participation. The easiness (first and second question) was measured using a likert scale, being (1) very difficult and (5) very easy. Likewise, the equality (third question) was measured using a likert scale, being (1) completely unequal and (5) completely equal. A Pre-test was applied, to observe educators' perceptions in the design activity. Before the session started, educators answered the questionnaire. During the demonstration, educators observed how a group of students interacted with the MTS. When the demo finished, a summary with the students' contribution was presented to educators in a web application. At the end, educators answered a post-test questionnaire, in order to explore their perception when using the MTS.

Experiment 2: Fifteen students were invited to test the usefulness of the MTS. Students conformed groups of 4–5 members to work collaboratively. Before the educator assigned a task to be designed in the MTS, the research group explained a general introduction of the use of the MTS and how to perform main actions. Then, for each student, a usability test was applied, which consists in the execution of twelve tasks (e.g. add an entity, edit an entity, add an attribute, etc.) in the MTS. This test included three variables: easiness to perform a task (likert scale, (1) very difficult and (5) very easy); perceived satisfaction (likert scale, (1) unsatisfactory—(5) very satisfactory) and perceived usefulness (likert scale, (1) useless—(5) very useful) of the MTS. In addition, students took the time used to execute all tasks.

4.2 Experimental Results

- **Experiment 1**: Wilcoxon signed-rank test resulted in significant differences for the three measured variables. Table 1 summarizes the descriptive statistics with the corresponding z and p values.
- **Experiment 2**: Results showed that 91 % of students (mode: 5, SD:0.62, median: 5) agreed that the MTS was easy or very easy to use. Also, all students reported positive responses (mode: 4, SD:0.49, median: 4) about satisfaction, meaning that none of them feel frustrated when using the MTS. As for the usefulness, most of the students (mode: 4, SD:0.70, median: 4) perceived that MTS was useful (50 %) and very useful (40 %). In addition, an average time of 13 min was used to perform all the tasks.

Table 1 Statistics for educator's perceptions

Variable	Median pre-test	Median post-test	z-value	p-value
Easiness to grade individuals	2	5	−2.859	0.004
Easiness to grade groups	4	5	−2.333	0.020
Equality of participation	2	4	−2.372	0.018

5 Discussion and Future Work

This paper aims to describe the implementation of a low-cost multi-tabletop system to support collaborative design assessment. Collaborative design activities are difficult to assess due to the hard effort that educator should put when observing each student while performing collaboration activities. Furthermore, it is desirable to have a tabletop system that would help to assess quantitatively these activities by showing a report about the actions performed with the MTS per student and an overall performance per group. Thus, twoexperiments were executed to validate the proposed system in terms of easinessof grading and equality of individual participation.

The presented approach lowers considerably the costs and the portability among other solutions. Non-specialized hardware was used in the implementation of the system. A detailed process for implementing the MTS was described to serve as a baseline for similar systems.

It has been proved from experiment with students that the proposed MTS is effective and useful for tracking individual contributions of students who are involved in collaborative work of data design. Students perceived that the use of MTS is more desirable than traditional approach (e.g. pen and paper based, whiteboard). Most of students expressed that it was not difficult to perform common actions in the MTS, meaning that the proposed design fits the needs of users.

In addition, the perception of educators about fairness for grading individuals indicated that, even though the presented work is a first prototype, it has a great potential to help educators giving an on-time feedback. Results showed that other means of collaboration in a work group do not support the design assessment, which in contrast, the proposed MTS gives the proper information.

Although, the contribution of a student is measured quantitatively, the quality of the contribution should be added to the contribution in order to have a more reliable and realistic measure. Quality can be measured by performing a deeper analysis of the student's interventions while doing the collaborative design; which implies a multimodal analysis of speech, gestures and input's actions to the system.

It would be interesting to consider the collaborative surfaces for design activities that are involved in software development. For instance, the development of UML diagrams, flowcharts, BPMN diagrams, among others. Also, in other engineering design activities, such as mechanical or industrial design.

References

1. Kaplan: Graduate recruitment report: employer perspectives. Technical report, Kaplan (2014)
2. National Association of Colleges and Employers: the skills and qualities employers want in their class of 2013 recruits. Technical report (2013)
3. Rick, J., Marshall, P., Yuill, N.: Beyond one-size-fits-all: how interactive tabletops support collaborative learning. In: Proceedings of the 10th International Conference on Interaction Design and Children pp. 109–117. New York, NY, USA, ACM (2011)
4. Martínez Maldonado, R., Kay, J., Yacef, K., Schwendimann, B.: An interactive teacher's dashboard for monitoring groups in a multi-tabletop learning environment. Intelligent Tutoring Systems. Lecture Notes in Computer Science, vol. 7315, pp. 482–492. Springer Berlin (2012)
5. Dillenbourg, P., Evans, M.: Interactive tabletops in education. Int. J. Comput. Support. Collab. Learn. **6**(4), 491–514 (2011)
6. Hilliges, O., Terrenghi, L., Boring, S., Kim, D., Richter, H., Butz, A.: Designing for collaborative creative problem solving. In: Proceedings of the 6th ACM SIGCHI Conference on Creativity & Cognition, pp. 137–146. New York, NY, USA, ACM (2007)
7. Geyer, F., Pfeil, U., Höchtl, A., Budzinski, J., Reiterer, H.: Designing reality-based interfaces for creative group work. In: Proceedings of the 8th ACM Conference on Creativity and Cognition, pp. 165–174. New York, NY, USA, ACM (2011)
8. Chory-Assad, R.M.: Classroom justice: perceptions of fairness as a predictor of student motivation, learning, and aggression. Commun. Q. **50**(1), 58–77 (2002)
9. Higgins, S., Mercier, E., Burd, E., Hatch, A.: Multi-touch tables and the relationship with collaborative classroom pedagogies: a synthetic review. Int. J. Comput. Support. Collab. Learn. **6**(4), 515–538 (2011)
10. Mercier, E.M., Higgins, S.E.: Collaborative learning with multi-touch technology: developing adaptive expertise. Learn. Instr. **25**, 13–23 (2013)
11. Tyng, K.S., Zaman, H.B., Ahmad, A.: Visual application in multi-touch tabletop for mathematics learning: a preliminary study. In: Visual Informatics: Sustaining Research and Innovations. Lecture Notes in Computer Science, vol. 7066, pp. 319–328. Springer, Berlin (2011)
12. Martínez, R., Collins, A., Kay, J., Yacef, K.: Who did what? who said that?: collaid: an environment for capturing traces of collaborative learning at the tabletop. In: Proceedings of the ACM International Conference on Interactive Tabletops and Surfaces, pp. 172–181. New York, NY, USA (2011)
13. Wellner, P.: Interacting with paper on the digitaldesk. Commun. ACM **36**(7), 87–96 (1993)
14. Cutler, L.D., Fröhlich, B., Hanrahan, P.: Two-handed direct manipulation on the responsive workbench. In: SI3D '97: Proceedings of the 1997 Symposium on Interactive 3D Graphics, ACM Press (1997)
15. Buxton, W., Fitzmaurice, G.W., Balakrishnan, R., Kurtenbach, G.: Large displays in automotive design. IEEE Comput. Graph. Appl. **20**(4), 68–75 (2000)
16. Stahl, O., Wallberg, A., Söderberg, J., Humble, J., Fahlén, L.E., Bullock, A., Lundberg, J.: Information exploration using the pond. In: Proceedings of the 4th International Conference on Collaborative Virtual Environments. CVE '02, pp. 72–79. New York, NY, USA, ACM (2002)
17. Jones, A., Moulin, C., Barthes, J., Lenne, D., Kendira, A., Gidel, T.: Personal assistant agents and multi-agent middleware for CSCW. In: IEEE 16th International Conference on Computer Supported Cooperative Work in Design, pp. 72–79 (2012)
18. Sinmai, K., Andras, P.: Mapping on surfaces: Supporting collaborative work using interactive tabletop. In: Baloian, N., Burstein, F., Ogata, H., Santoro, F., Zurita, G. (eds.) Collaboration and Technology. Lecture Notes in Computer Science. vol. 8658. pp. 319–334, Springer International Publishing (2014)
19. Shen, C., Ryall, K., Forlines, C., Esenther, A., Vernier, F.D., Everitt, K., Wu, M., Wigdor, D., Morris, M.R., Hancock, M., Tse, E.: Informing the design of directtouch tabletops. IEEE Comput. Graph. Appl. **26**(5), 36–46 (2006)

20. Xambó, A., Hornecker, E., Marshall, P., Jordà, S., Dobbyn, C., Laney, R.: Let's jam the reactable: peer learning during musical improvisation with a tabletop tangible interface. ACM Trans. Comput.-Hum. Interact. **20**(6):36:1–36:34 (2013)
21. Martínez Maldonado, R., Dimitriadis, Y., Kay, J., Yacef, K., Edbauer, M.T.: Orchestrating a multi-tabletop classroom: from activity design to enactment and reection. In: Proceedings of the 2012 ACM International Conference on Interactive Tabletops and Surfaces, pp. 119–128. New York, NY, USA, ACM (2012)
22. Kaltenbrunner, M., Bovermann, T., Bencina, R., Costanza, E.: Tuio a protocol for table-top tangible user interfaces (2005)
23. Laufs, U., Ruff, C., Zibuschka, J.: Mt4j-a cross-platform multi-touch development framework. arXiv preprint arXiv:1012.0467 (2010)
24. Paulson, B., Hammond, T.: Paleosketch: accurate primitive sketch recognition and beautification. In Proceedings of the 13th International Conference on Intelligent User Interfaces, pp. 1–10. ACM (2008)

Creativity as a Key Ingredient of Information Systems

Vítor Santos, Jorge Pereira, José Martins, Ramiro Gonçalves and Frederico Branco

Abstract Resorting to creativity technique and their use to help innovation in the area of information systems had a growing interest. In fact, the global competitiveness and the organizations ability to make effective use of information technology and to focus on innovation and creativity are recognized as being important. So, the perspective of using creativity techniques seems to be promising. In this research work we argue that is possible in all IS areas to take advantages of the use of creative processes. We give a pragmatic reasoning and examples for the introduction of creative processes in all the main IS areas.

Keywords Information systems · Creative thinking · Innovation

---

V. Santos
NOVA IMS—Information Management School, Universidade Nova de Lisboa, Lisbon, Portugal
e-mail: vsantos@novaims.unl.pt

V. Santos
MagIC, Nova University of Lisbon, Lisbon, Portugal

J. Pereira · J. Martins · R. Gonçalves · F. Branco
University of Trás-Os-Montes E Alto Douro, Vila Real, Portugal
e-mail: jorge.m.g.pereira@gmail.com

J. Martins
e-mail: jlbandeira@gmail.com

F. Branco
e-mail: fbranco@utad.pt

J. Martins · R. Gonçalves (✉) · F. Branco
INESC TEC and UTAD, Porto, Portugal
e-mail: ramiro@utad.pt

J. Mejia et al. (eds.), *Trends and Applications in Software Engineering*, Advances in Intelligent Systems and Computing 405,
DOI 10.1007/978-3-319-26285-7_24

1 Introduction

The capacity of enterprises effectively using information technologies and betting on innovation and creativity is recognized as one important factor on the competiveness and agility of the enterprises [1–3]. These take natural benefits through creativity and innovation by restructuring their processes, projects and products [4].

Find the best ways to take advantage of Information Systems (IS) is a vital for the success and competitiveness of companies [3, 5]. The diversity of the sectors in corporate activity, the different contexts and organizational structures are, along with the growing complexity of the globalized world of business, a huge challenge for the effectiveness of this goal.

In this context, the chances of resourcing to known creativity techniques or their adaptations, in order to mediate the spawning of ideas, help produce new combinations, supply unexpected answers, as well as original, useful and satisfactory, in the area of Information Systems is challenging.

In this article we defend that is possible to take advantages of the use of creative processes in all IS areas.

2 Research on Creativity

In the last 60 years research on creativity has been vast. Creativity and creative processes are the object of case studies in many fields of knowledge, namely in psychology, cognitive sciences, neuro-biology, education, philosophy, theology, technologies, sociology, linguistics, management, innovation, sciences, economy, among others [6].

Psychology and cognitive sciences have focused their attention on the study of mental representations and the underlying creative thinking. According to Candeias [7] the main focus until the 1970s was on approaches related to creativity based on personality studies in order to identify the creativity features in different domains. From then on the main focus of research changed to the components of creative thinking and the resolution of problems [7].

The study of creativity in neuro-biology has had—in the last few years—a reasonable success with, for example, the appearance of works that attempt to link individual creativity to communication between areas of the brain that are not normally linked together [8]. Also in epistemology and theology is research that attempts to shed light on creativity. Philosophy tries to answer questions such as: What is creativity? How does it come to be? How does creativity manifest itself in findings, inventions, science and art? What is the role that creativity has in the construction of the subject? Theologians debate the connection between creativity and holy and divine inspiration, attempting to solve the tension between human creation vs new creation as an expression of God's work.

Also in the fields of Sociology and Education it is possible to identify a renewed interest in the issue of creativity. Recently, researchers have turned their attention to the introduction of strategies in the classroom that would allow for the stimulation and development of creativity among students. Creativity is also growingly encouraged as a means to promote both autonomous learning and increased interaction between teachers and students [9].

In the fields of management, innovation, entrepreneurship, economics and technologies the importance of creativity, as a first step for the birth of inventions and innovation has had a strong and diversified focus in virtually every domain. This is particularly the case of information systems and technology.

3 Creativity in Information Systems

It is common sense that large organizations are more likely to be bureaucratic and complex, while the smallest are more agile and flexible [10]. Although creativity and innovation are certainly important to all organizations, it is also assumed that much of the success to be creative and innovative depends on the conditions and market opportunities [11]. There seems to be some truth in the observation that small businesses can react faster to changes in their environment and can easily reorganize itself to take advantage of changes in the surrounding markets [5, 12].

This flexibility allows often take advantage of creativity and innovation. The innovation is often mentioned as a consequence of the creativity [13–15] and, therefore, creativity is seen as the key to organizational survival [10].

Dhillon et al. [10] consider that the papers published in this area have revealed that the Information Systems research community has been successful in incorporating existing proposals and the development of new and interesting research area.

While many organizations are creative and innovative in part because of the need to remain competitive, the effective use of creativity and innovation in information systems and technologies and still scarce and is a major challenge.

As argued by Martins et al. [16] in order to ensure a valid and competitive existence, organizations are adoption information systems and technologies, in particular those who require an increased adaptability and creativity, thus fostering the relations with their customers.

Find the best way to combine the creativity, flexible and divergent thinking with analytical thinking, pragmatic and rigid, essential to the areas of technology and information systems, and have significant gains is not immediate or trivial.

So, establish a connection between creativity competitive advantages in an organization and define the roles of Information and Communication Technologies and Information Systems in this connection still are open questions.

In an attempt to address these issues, a wide range of theoretical and practical approaches have been used by several authors [10, 17–24].

The use of creativity can also be used to re-think and innovate in how employees and information professionals can use the information systems and technologies so that they can turn more productive [15].

Managers should try to determine their self-organizing capabilities and provide employees the opportunity to use their creative abilities to propose new activities for the organization, which will optimize processes and improve the relationship between people. For example, looking for innovative ways to use information technologies, such as email, forums, intranets, social networks, among others.

Creativity has been advocated as central to the development of Information Systems [17, 23, 25–28]. Higgins et al. [26] argues that creativity plays an important role in all aspects of the development of information systems, from design, requirements analysis, design, to programming.

3.1 *Creativity in Information Systems Development*

According Varajão [29], we can divide the Information Systems Development (ISD) activity in five phases: Analysis, Design, Construction, Implementation and Maintenance. Figure 1 illustrates the main ISD phases and activities.

The systems analysis is to define the problem, identify its causes, specify the solution and identify the information requirements that must be met in accordance with the organization's expectations. The system requirements specify try in full the

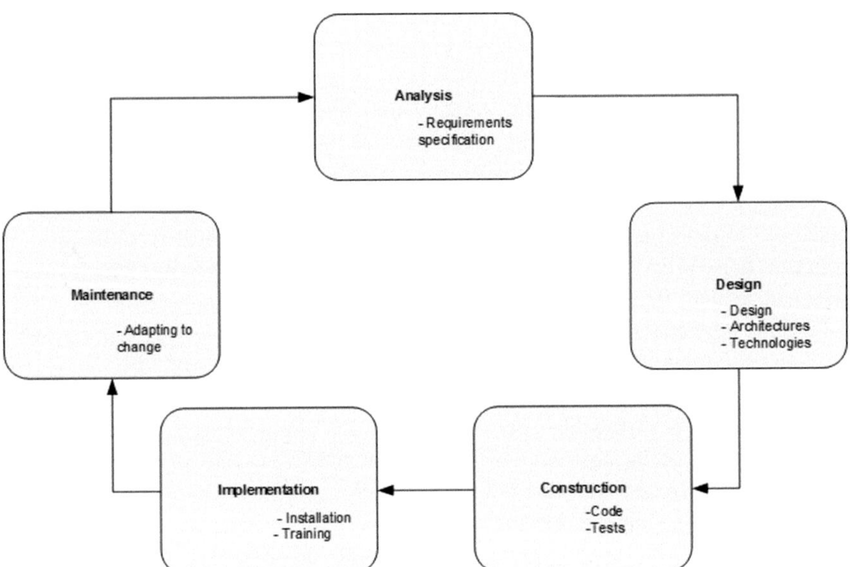

Fig. 1 Information systems development phases—adapted from Varajão [29]

necessary IS characteristics to meet the user information needs and the needs of the organization in general.

There are two types of requirements: the functional and non-functional. Functional requirements are those that define what the system has to be able to run. Non-functional requirements concerning general characteristics of the system and define the functions necessary for proper functioning of the system, for example, ergonomics performance, security and usability.

In this phase the use of creative processes allows you to tackle and find solutions to new problems and domains, unique contexts, new applications and recombine existing methods and techniques.

The systems Design activity is to specify in detail the functions that the system must ensure. Specifications can be of different types: interface specification, data specification, specification processes and technical environment specification.

The phase of the design of a computer system is undoubtedly one of the most interesting from the perspective of engineering and space for creativity. In this phase the system specification, resulting from the analysis phase must be converted into something that meets the requirements, and can be built, that is, produce the design and complete specification of the computer system, describing their data components, processes, communications and interface.

The Systems Construction is a technical activity for which the specifications are transformed into software and hardware. The tasks of acquisition and/or development of software and hardware, testing, prepare support documentation and the integration of all components into a functional information system are included in the systems construction.

Find the algorithms that allow to build the systems and translate them into efficient code requires very specific expertise and creativity in the design of new solutions.

Implementation is the process of putting the system to work within the organization so that it can be successfully used. Includes software and hardware installation tasks, preparation of facilities, train technical people and the conversion of existing systems to new systems.

Bringing new system into for use requires often to make changes (with greater or lesser depth) in the existing organizational structures and processes.

This can be caused by several factors, including the simplification of procedures introduced by the new system, the automation of manual processes, system integration, and the operating rigidity of the new system.

To minimize the impact of the implementation and accommodate the new system with maximum smoothness possible in the organization is necessary to promote a creative and open approach.

Finally, after the completion of the implementation follows the maintenance phase. Over time, due to the changing needs of users and the organization, changes in technology and detection of errors, the system will require updates, corrections, changes, adjustments and expansions, both hardware and software. The longevity of IS depend greatly on the quality and importance given to its maintenance. As the construction phase, find the algorithms that improve and adapt the systems and translate them into efficient code requires, in addition to technical knowledge, creative ability.

3.2 *Creativity in Information Systems Planning*

The role of Information Systems Planning has become crucial for the development of effective strategic plans in organizations [30, 31]. The increasing uncertainty in the markets has encouraged organizations to be more proactive. On the one hand, information technology provides a set of opportunities for gaining competitive advantage. This requires strategic alignment and a fit of Information Systems with the strategies, goals and operations of organizations. On the other hand, organizations acknowledge that the ability to provide a quick response to unforeseeable events is paramount for their survival [32].

The activity of information systems planning faces great challenges, since, on the one hand, fast technological developments make it hard to judge what the future holds [3]. On the other hand determining the best way to place technologies at the service of organizations is a difficult task, as environmental and contextual changes may make previous decisions rapidly obsolete.

The activity of information systems planning adds further difficulties, since, on the one hand, quick technological developments makes it hard to judge what the future holds and, on the other hand, to determine the best way to place technologies at the service of organizations, as anticipating changes in almost constant surroundings is a difficult activity. Managers are challenged, as never before. They must have the ability to perceive the multiplicity of changes that are taking place and the ramifications that they will have. As managers, they must predict the future and follow a course that leads a team or organization to a position where they can compete in an effective way and reach prosperity and sustainability. Thus, planning is considered to be one of the main activities of managers and its success is fundamental for the good performance of organizations.

The conception and study of information systems is complex in itself. Its complexity has its origin in the fact that information systems combine human and technological resources and—in a transverse way—they involve human activities, organizations, politics, markets, environment, industry, etc. ISP, as a planning activity, adds to the difficulty of trying to anticipate the constraints and also the opportunities that may occur in the future.

A problem in ISP is the absence of a sufficiently broad strategy to allow in an easy, flexible and effective way to introduce creative processes in different ISP approaches, thus stimulates the production of ideas, which produce new combinations, which obtain original and useful answers and, consequently, generate innovation in Information Systems and in the way that they are used. One example

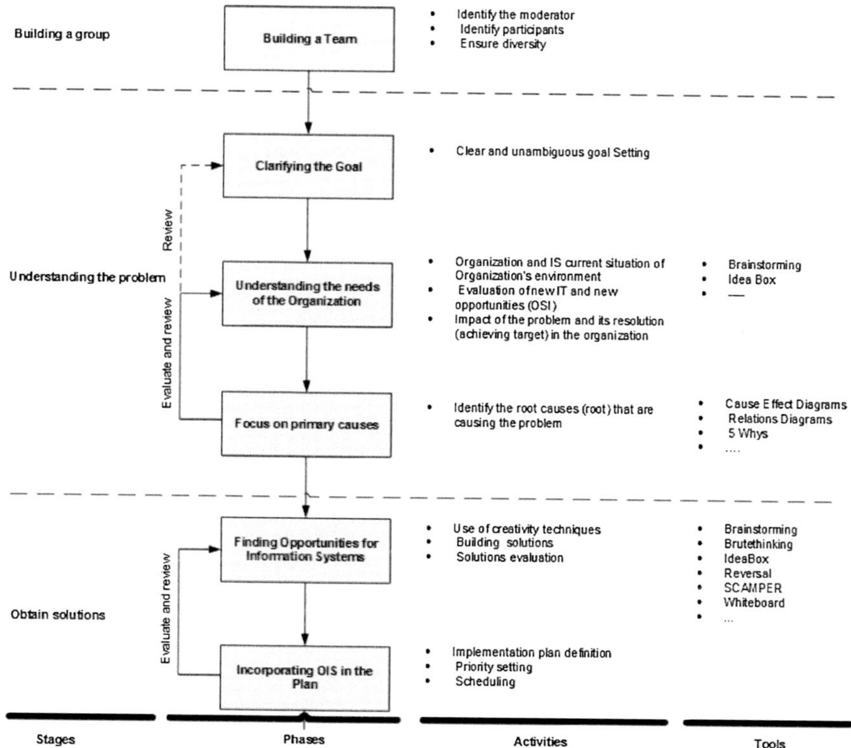

Fig. 2 Vision of the generic method for PSI problem solving

of a methodological proposal for the introduction of creativity in ISP is the "Creative Potentiation Method" [20]. Figure 2 illustrates the ISP is the creative potentiation method.

3.3 *Creativity in the Information Systems Exploration*

"Information Systems Exploration (ISE) is the activity responsible for the maintenance of the IS [29]. As stated in Varajão [29], the ISE can be divided into three main activities: Systems Operation, Human Resources Management and Information Technology Administration".

Ensure the effective operation of the systems, managing remove them as much value to the organization and tailor them to the daily needs implies management capacity but also flexibility and creativity to overcome the shortcomings and unforeseen difficulties.

4 Conclusions and Future Work

Creativity is important in all information systems activities, whether planning, management, development or exploitation. Enables find innovative solutions, improving existing systems, go around obstacles, reduce costs, use and reuse existing resources more efficiently, better align the systems with the business needs and take better advantage of individual contributions.

Of the performed analysis stands out the fact that all IS areas can take advantages from the use of creative processes, mainly because the flexibility achieved will increase the benefits associated with the adoption and use of the referred technologies.

In our opinion, we can trace a strategy that is consistent relative to the introduction of creativity processes and innovation in Information Systems, as well as the operationalization of its application through simple and practical methods.

The next stages to be introduced in this ongoing work are mainly related to the development of new theoretical methods that support the design and development of introduction of creativity in IS strategies, including improvement of the current methodologies, the refinement of validation criteria, the analysis of its applicability, evaluation, and as well as the implementation of eventual corrections.

References

1. Gonçalves, R., Barroso, J., Varajão, J., Bulas-Cruz, J.: Modelo de las iniciativas del comercio electrónico en organizaciones portuguesas. Interciencia: Revista de ciencia y tecnología de América **33**, 120–128 (2008)
2. Martins, J., Gonçalves, R., Pereira, J., Cota, M.: Iberia 2.0: A way to leverage Web 2.0 in organizations. In: 7th Iberian Conference on Information Systems and Technologies (CISTI 2012), pp. 1–7. Madrid, Spain (2012)
3. Pereira, J., Martins, J., Gonçalves, R., Santos, V.: CRUDI framework proposal: financial Industry Application. Behav. Inf. Technol. (2014)
4. Cooper, R.B.: Information technology development creativity: a case study of attempted radical change. Mis Quarterly 245–276 (2000)
5. Branco, F., Gonçalves, R., Martins, J., Cota, M.: Decision Support System for the Agri-food Sector—The Sousacamp Group Case. In: World Conference on Information Systems and Technologies. AISTI, Azores, Portugal (2015)
6. Tarrida, A., Femenia, D.: Dirigir la creatividad: Una aproximación al funcionamiento intelectual de los directores de cine. In: Morais, M.d.F., Bahia, S. (eds.) Criatividade: Conceito, Necessidades e Intervenção, Psiquilibrios Braga (2008)
7. Candeias, A.: Criatividade: Perspectiva integrativa sobre o conceito e a sua avaliação. Criatividade: Conceito, necessidades e intervenção, pp. 41–64 (2008)
8. Heilman, K., Nadeau, S., Beversdorf, D.: Creative innovation: possible brain mechanisms. Neurocase **9**, 369–379 (2003)
9. Moraes, M.d.F., Bahia, S.: Criatividade Psiquilibrios, Braga (2008)
10. Dhillon, G., Stahl, B., Baskerville, R.: Creativity and intelligence in small and medium sized enterprises: the role of information systems. Information Systems–Creativity and Innovation in Small and Medium-Sized Enterprises, pp. 1–9. Springer (2009)

11. Tidd, J., Bessant, J.: Gestão da inovação-5. Bookman Editora (2015)
12. Gonçalves, R., Martins, J., Pereira, J., Oliveira, M., Ferreira, J.: Enterprise Web Accessibility Levels Amongst the Forbes 250: Where Art Thou O Virtuous Leader? J. Bus. Ethics **113**, 363–375 (2013)
13. De Bono, E.: Lateral Thinking for Managers. Penguin Books, London (1990)
14. Hurson, T.: Think Better: An Innovator's Guide to Productive Thinking. McGraw Hill Professional, New York (2012)
15. Oliveira, J.: Gestão da inovação. Sociedade Portuguesa de Inovação. Principia, p. 50 (1999)
16. Martins, J., Gonçalves, R., Oliveira, T., Pereira, J., Cota, M.: Social networks sites adoption at firm level: a literature review. In: CISTI'2014—Iberian Conference on Information Systems and Technologies, Barcelona, Espanha (2014)
17. Conboy, K., Wang, X., Fitzgerald, B.: Creativity in Agile Systems Development: a literature review. Information Systems–Creativity and Innovation in Small and Medium-Sized Enterprises, pp. 122–134. Springer (2009)
18. Kotzé, P., Wong, W., Jorge, J., Dix, A., Silva, P.: Creativity and HCI: from experience to design in education: selected contributions from HCIEd 2007, 29–30 March 2007, Springer, Aveiro, Portugal (2008)
19. Ellis, C., Gibbs, S., Rein, G.: Groupware: some issues and experiences. Commun. ACM **34**, 39–58 (1991)
20. Santos, V., Amaral, L., Mamede, H., Gonçalves, R.: Creativity in the information systems planning process. In: IGI-Global (ed.) Handbook of Research on Innovations in Information Retrieval, Analysis, and Management, (2015)
21. Mamede, H., Santos, V.: Architecture for a creative information system. Information Systems–Creativity and Innovation in Small and Medium-Sized Enterprises, pp. 113–121. Springer (2009)
22. Murthy, U.: Conducting creativity brainstorming sessions in small and medium-sized enterprises using computer-mediated communication tools. Information Systems–Creativity and Innovation in Small and Medium-Sized Enterprises, pp. 42–59. Springer (2009)
23. Shneiderman, B., Fischer, G., Czerwinski, M., Resnick, M., Myers, B., Candy, L., Edmonds, E., Eisenberg, M., Giaccardi, E., Hewett, T.: Creativity support tools: report from a US National Science Foundation sponsored workshop. Int. J. Hum Comput Interact. **20**, 61–77 (2006)
24. Pissarra, J.: Geração de ideias mediadas por computador. Universidade Lusiada Editora, Lisboa (2009)
25. Carayannis, E., Coleman, J.: Creative system design methodologies: the case of complex technical systems. Technovation **25**, 831–840 (2005)
26. Higgins, L., Couger, J., McIntyre, S.: Creative approaches to development of marketing information systems. In: Proceedings of the Twenty-Third Annual Hawaii International Conference on System Sciences, vol. 4, pp. 398–404. IEEE (1990)
27. Gallivan, M.: The influence of system developers' creative style on their attitudes toward and assimilation of a software process innovation. In: Proceedings of the Thirty-First Hawaii International Conference on System Sciences, vol. 6, pp. 435–444. IEEE (1998)
28. Lobert, B., Dologite, D.: Measuring creativity of information system ideas: an exploratory investigation. In: Proceedings of the Twenty-Seventh Hawaii International Conference on System Sciences, vol. 4, pp. 392–402. IEEE (1994)
29. Varajão, J.: A arquitectura da gestão de sistemas de informação. FCA (1998)
30. Chen, D., Mocker, M., Preston, D., Teubner, A.: Information systems strategy: reconceptualization, measurement, and implications. MIS Q. **34**, 233–259 (2010)
31. Lederer, A., Sethi, V.: Critical dimensions of strategic information systems planning. Decis. Sci. **22**, 104–119 (1991)
32. Allaire, Y., Firsirotu, M.: Coping with strategic uncertainty. MIT Sloan Manage. Rev. **30**, 7 (1989)

Part V
Organizational Models, Standards and Methodologies, Processes in Non-software Domains

Decreasing Rework in Video Games Development from a Software Engineering Perspective

Hugo A. Mitre-Hernández, Carlos Lara-Alvarez, Mario González-Salazar and Diego Martín

Abstract Video game industry is becoming increasingly important due to its revenues and growing capabilities. Information complexity and process agility are limitations for developing a videogame and they may lead to rework. Many rework problems are related to unspecified or ambiguous requirements in game design. For reducing rework, this article proposes an agile development process for video games that aligns the Scrum instance of the software development Project Pattern (sdPP) and the improved Game Design Document (iGDD). For measuring the rework induced by different alternatives, we conducted a case study that compares the proposed approach against a conventional counter proposal in game industry; the results prove that our proposal generates less normalized rework than the counter proposal.

Keywords Rework · Video games · Software engineering · Requirements engineering

H.A. Mitre-Hernández (✉) · C. Lara-Alvarez · M. González-Salazar
Computer Science Department, Center for Research in Mathematics (CIMAT),
Av. Universidad 222, 98068 Zacatecas, Mexico
e-mail: hmitre@cimat.mx

C. Lara-Alvarez
e-mail: carlos.lara@cimat.mx

M. González-Salazar
e-mail: remylebv@cimat.mx

D. Martín
Telematics Department, Technical University of Madrid, Av. Complutense 30,
28040 Madrid, Spain
e-mail: diego.martin.de.andres@upm.es

J. Mejia et al. (eds.), *Trends and Applications in Software Engineering*,
Advances in Intelligent Systems and Computing 405,
DOI 10.1007/978-3-319-26285-7_25

1 Introduction

Video games are important economically, as innovative leaders and as an alternative to solve issues outside the entertainment area; they constitute the main entertainment industry, with continuous growth since their appearance and billions of dollars in sales and revenues [1]. Video games have proven useful outside the entertainment area with good results for solving problems, training, diagnosing, predicting, and teaching among others [2].

In the video game industry, the development process is commonly composed by three stages: pre-production, production and post-production [3]. The *pre-production* stage focuses mainly on creating the game concept and design. The *production* stage creates and validates the software; this stage also produces sounds and music required by the game. Finally, the *post-production* stage distributes and maintains the game; it also manages the feedback coming from different sources—i.e. specialized video game reviewers–.

Some games are complex systems requiring significant effort in the first two stages; these complexities can increase the amount of rework and consequently, the cost of the game. *Rework* is defined as any additional effort required for finding and fixing problems after documents and code are formally signed-off as part of configuration management [4]. Thus, end-phase verification and validation are usually excluded, but debugging effort during integration and system testing is included. To compare different products, rework effort is sometimes "normalized" by being calculated as a percentage of development effort [4]; the rework is generally considered to be potentially avoidable work that is triggered to correct problems.

Rework issues in game development usually start at the pre-production stage. It has been reported that 65 % of problems in game development are generated at this stage and are related to unspecified or ambiguous requirements in game design [5, 6]. At the production stage, developers can ask the game designer for clarification of the missing information or can make their best assumption. In either case, a rework is done; even, rework can rise when wrong assumptions about the requirements are made. Hence, we need a way to describe the game design with enough formality and detail, a Software Requirements Specification (SRS) document can be used for designing the software of the game [7, 8]. In sum, a more formal requirement specification is needed to avoid rework problems in the game development process.

The game *design* is central in game development; this activity transforms an idea into a detailed description of the game. Formally, the game design is the process of imagining a game, defining the way it works, describing the elements that make up the game (conceptual, functional, artistic, and others), and transmitting that information to the team that will build the game [9]; this information is reflected in the *Game Design Document* (GDD).

We consider that the rework can be substantially reduced in the pre-production stage. The approach proposed in this article is composed by (i) a project pattern adapted to game agile development, and (ii) an improved GDD [10] based on requirements engineering. Software process patterns encapsulate the solution for a specific development project; therefore, it is possible to adapt them to the game development process. On the other hand, the requirement engineering is helpful because it offers characteristics such as correctness, unambiguously, completeness, consistency, stability, verifiability, modifiability, traceability [11] and structurability [12].

The rest of this article is organized as follows: Sect. 2, summarizes the game development methodologies directly related to our approach; Sect. 3, introduces the proposed approach; Sect. 4 compares empirically our approach against a classical methodology for game development in terms of rework; Sect. 5, discusses the results; and finally, Sect. 6 concludes this work.

2 Related Work

As stated earlier, our proposal aligns a project pattern with the improved GDD; the following paragraphs overview the work related to these two components.

2.1 Development Project Pattern

The videogame development is a form of software development that adds additional requirements, e.g., artistic aspects; hence, many of the management tools and standards from the software industry can be useful for game development. Game projects are usually more complicated than software projects because they involve a multidisciplinary team and they usually have more uncertainty around project goals.

Software development models—e.g., waterfall, iterative, or extreme—can be used for developing videogames [3]. In general, the waterfall model is considered inadequate because it is highly structured and it cannot be adapted to changes in the requirements; therefore, more flexible models are needed [13–15].

Agile methodologies—i.e. Scrum [16] or eXtreme Programming [17]—are better suited for the challenges of game development [18, 19]; they have been adapted to game development by using other tools as complements: user stories [18], game design documentation [20], or workshops for strengthening the interaction between clients and developers [21].

Patterns [22] are used to solve a generic problem: given a narrative and context of the problem to be solved, they propose a solution. They can be used for formalizing the knowledge about the development process; [23] proposes the *Software Development Project Pattern* (sdPP) framework. For testing this approach [23], generates four instances of the sdPP with agile development models; one of these instances—Scrum sdPP—is suitable for game agile development because it allows to follow an iterative process without sacrificing creativity. The resulting workflow and productflow can guide game developers between the activities and their corresponding input and output products.

Table 1 Description and characteristics of sections of the iGDD [10]

iGDD section	Description	SRS characteristics
Overview	Describes briefly the most important aspects of the game	Relations with other documents, and common language for better understanding
Mechanics	Describes the elements of the game	Organization of game requirements (objects organization)
Dynamics	Describes how the elements of the game will take action in the game	Organization of game requirements. Relation of complexity with gamer profile
Aesthetics	Describes what the player perceives directly through their sense, like what he sees and hears	–
Experience	Highlights important aspects of the game and what you hope to achieve from these aspects	Decision making based on tradeoffs of game parts. Quality attributes on video games
Assumptions and constraints	Narrates the aspects of the design assumptions and limitations of the game, either technical or business	Knowledge of game parts for reviews. Limitations or boundaries of video game

2.2 Improved Game Design Document

The *Mechanics, Dynamics and Aesthetics* (MDA) framework [13] is an iterative approach for designing and tuning video games; it first defines the aesthetics, then the dynamics—trying to fulfill aesthetics–, and finally the game elements—that bring the required interaction–; this framework offers an interesting way of designing and tuning games, but it does not provide the tools or methods to construct the details of such mechanics, dynamics and aesthetics. On the other hand, several approaches suggest to use different abstraction levels for the game design documentation, but they are too specific for certain types of games [24, 25].

The *Game Design Document* (GDD) plays a key role for every game project, i.e., a poorly elaborated GDD can lead to rework and loss of investment in production and postproduction phases; to address these issues, in [10] we propose the *improved GDD* (iGDD). The iGDD has different abstraction levels [9]; it is based on the Taylor's GDD template [24], but it incorporates the (MDA) framework and the best practices from Software Requirements Specification (SRS) [7, 8]; sections of the iGDD are shown in Table 1.

3 Proposed Approach

The sdPP model is defined as a problem–solution pair; in our case, the solution is closely related to agile game development with Scrum and the use of the iGDD into the development process. Proposed improvements to the Scrum sdPP are: (i) the alignment of activities described in the sdPP's workflow and productflow to the iGDD; and (ii) the incorporation of the iGDD sections to the Scrum sdPP as products. For adapting Scrum sdPP to game developing new activities were added while others were modified; some of these activities are associated to iGDD sections (products) as shown in Fig. 1. The modified (or added) activities and their associated iGDD products are described in the following paragraphs:

- **Create overview (product: *overview* of the iGDD** [10]), describe the game in a brief abstract, identify the main objectives of the game, the genre of the game, ask questions—e.g., why the game is worth doing–, define which type of players would like to play the game, and what will be the main activities that the player will be doing while playing the game.
- **Design High level game (product: *overview* of the iGDD** [10]), define some main features of the game: the game modalities (single player, multiplayer, online, arcade mode, history mode, among others), the platform or platforms on which the game is intended to run, the game theme (medieval, futuristic, western, among others), the game story and an initial scope of the levels, size and time that the game may require.
- **Design high level game architecture (product: *assumptions and constraints* of the iGDD** [10], Table 1). Review the technical settings to modify the *assumptions and constraints* and determinate any technical constraint that the game may have. *Technical settings* include: the standards, conventions, technology, resources and architecture selected for the game.
- **Design Game (products: *mechanics, dynamics and aesthetics* of the iGDD** [10], Table 1). The design of a level of the game will involve all three categories, while the design of the main character and all his action will involve only mechanics and aesthetics.

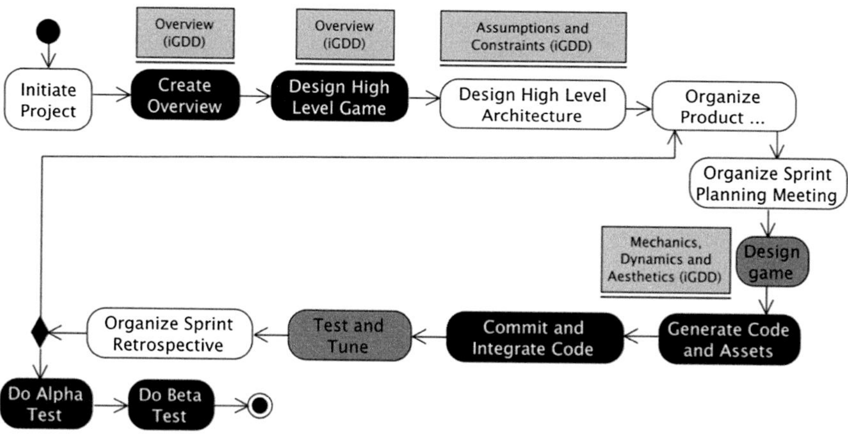

Fig. 1 Productflow of the scrum sdPP adapted to agile video game development. The added activities are in *black* and the modified ones are in *gray*; these activities are associated to iGDD sections (products) in *light gray*

- **Generate code and asset**. Create game elements, as suggested by Keith [18]. These elements include code and assets, i.e., music or animations.
- **Commit and integrate code**. Integrate game elements, as suggested by the Agile Alliance [26]; a totally integrated game allow to have "at any time" a version suitable for release.
- **Do test and tune**. Test the resulting product of the sprint, as suggested by Keith [18]; small adjustments can be made to polish the game, but radical changes should be placed in the product backlog to consider them in the next sprint. The result from this activity will be a potentially shippable product.
- **Do alpha and beta tests**. Find and remove bugs in alpha test [9, 18, 27], and test the experiences of the possible market that will play the game in the beta test.

4 Research Process

The research process is based on guidelines proposed in [28] for empirical evaluation in software engineering.

4.1 Plan

The objective of the case study is to measure whether or not our approach helps to decrease rework while developing a game. For measuring the rework, any artifact put to test for the first time starts to register rework time after the test is done.

Twelve junior software engineers at the Center for Research in Mathematics carried out this empirical study; these engineers (hereafter, participants) were students of the videogame course in the software engineering master degree program. For our study two groups were considered:

Group A: uses the approach proposed in this article;
Group B: uses the Taylor's GDD [24] and the agile game development with Scrum [18].

4.2 Execution

The following activities were carried out:

1. **Team Creation**. The professor interviewed the students for knowing their experience in similar projects and created two groups as homogeneous as possible.

2. **Team Training**. The researcher trained the teams in the required design documents and development methods.
3. **Conduct Experiment and Data Collection**. The researcher conducted a sprint planning meeting with each team. In this meeting the base requirements and its priorities were presented. The team divided the base requirements into user stories for planning the sprint. Then, the teams worked on the user stories of the sprint for 2 weeks recording the time spent on each user story. The researcher conducted a sprint end meeting where the user stories were classified as completed, unfinished or rejected, and the review of the correct use of objects (GDD and game development model) used by teams during the sprint.

Finally, each team and researcher planned the date for the next sprint planning meeting, ensuring that the time difference between sprint end and sprint planning was less than 5 days. This process continued until the base requirements were finished or the time available ended. The time for these steps were 3 months, where the teams should have from four to five sprints and finish from ten to fifteen base requirements.

4.3 *Results*

The data of times and user stories associated with base requirements collected during the case study were recorded in the log of Kanban web. The data was validated at the end of each sprint planning.

The Wilcoxon signed-rank test was used to compare the rework of groups A and B.

As shown in Fig. 2, the normalized mean of rework for group A was 2.73 %, while for group B was 11.50 %. A Wilcoxon signed-rank test showed that the

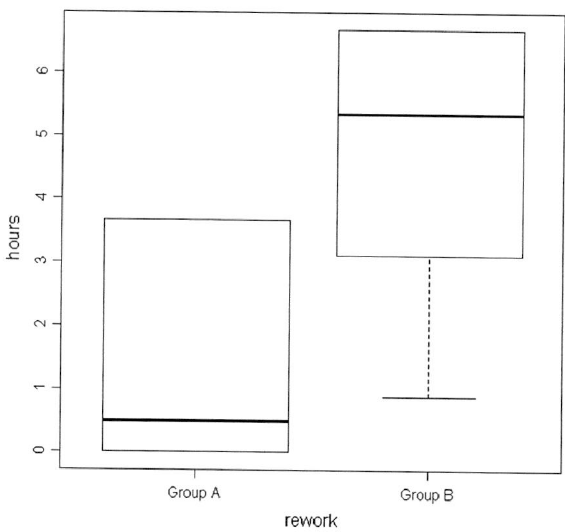

Fig. 2 Productflow comparison of rework between groups A and B

iGDD + sdPP technique induces significantly less rework than the rework induced in the group that used GDD + Agile GD with Scrum ($p = 0.0085$). *This proves that our proposal generates less rework than the counter proposal.*

5 Discussions

We observe the following benefits of the iGDD for reducing rework:

- By defining the overview section of the iGDD, participants clarify the information of the objectives, justification, gameplay, game features and player characteristics. This section allows to easily interpreting the requirements in the following stages.
- By defining the assumptions and constrains of the iGDD, participants discover the context and limits of the high level architecture. In terms of requirements engineering, we learned that limitations and boundaries clearly defined can lead to a feasible development.
- By defining the mechanics and dynamics of the iGDD, participants obtain: (i) well organized game elements, and (ii) good alignment of complexity with gamer profile. The traceability of game elements facilitated participants to reduce rework in the development of game elements.

Our main observation to the adapted Scrum sdPP was the traceability of the activities to the iGDD sections, allowing participants to retrieve information about the game elements at any time during process execution.

Summarizing the opinions of the participants, the adapted sdPP is clearer, more detailed, concise, and accurate than the conventional approach. Moreover, the documentation was considered better structured, and it requires less effort in interpretation and design.

The main limitation in the case study is that the teams didn't have the time for creating game elements—e.g., the sound part of the game–; hence, external sources were used to complete the game. Another observed limitation is that it is hard to coordinate the scrum meeting when the members of a team have different schedules.

6 Conclusions and Further Work

This article describes how software development patterns make easer the use of agile game development process; the proposed approach is composed by an agile development process—an adapted Scrum instance of the sdPP—and an improved game design document iGDD that takes advantage of the requirements engineering perspective.

A case study was conducted, validating that our proposal generates less rework than a conventional counter proposal. Finally, aspects that contributed to the reduction of rework with the use of the iGDD and the sdPP's instance were discussed.

We consider that the proposed approach not only reduces the rework but also can give better quality products that enhance the user experience. On the other hand, a management tool can be used for increasing the productivity of medium or large scale game projects. We are studying empirically both the quality and productivity induced by this approach.

Acknowledgments This research was partially funded by the National Council of Science and Technology of Mexico (CONACyT) through the project "Strengthening of the master of software engineering program with the integration the research line in Human-Computer Interaction" (ZAC-2013-C04- 226098) and the project "Optimization of industrial processes based on simulators, interfaces and software assurance" (CATEDRAS-3163).

References

1. Essential facts about the computer and video game industry (2014)
2. Jason: Gaming is good for you (infographic), http://www.affordableschoolsonline.com/gaming-is-good-for-you-infographic/
3. Bethke, E.: Game Development and Production, Pap/Cdr edition. edn. Wordware Publishing Inc., Plano (2002)
4. Kitchenham, B., Pfleeger, S.L.: Software quality: the elusive target. IEEE Softw, vol. 13, pp. 12–21. (1996)
5. Petrillo, F., Pimenta, M., Trindade, F., Dietrich, C.: What went wrong? A survey of problems in game development. Comput. Entertain. **7**, 1–22 (2009)
6. Petrillo, F., Pimenta, M., Trindade, F., Dietrich, C.: Houston, we have a problem…: a survey of actual problems in computer games development. In: ACM Symposium on Applied Computing. pp. 707–711. ACM (2008)
7. Callele, D., Neufeld, E., Schneider, K.: Requirements engineering and the creative process in the video game industry. In: Proceedings. 13th IEEE International Conference on Requirements Engineering, pp. 240–250. IEEE (2005)
8. Callele, D., Neufeld, E., Schneider, K.: A report on select research opportunities in requirements engineering for videogame development. In: The 4th international workshop on Multimedia and Enjoyable Requirements Engineering, pp. 26–33. (2011)
9. Rollings, A., Adams, E.: Andrew Rollings and Ernest Adams on Game Design. New Riders, 1st edn. (2003)
10. Gonzalez, M., Mitre, H.A., Lemus, C., Gonzalez, J.L.: Proposal of game design document from software engineering requirements perspective. In: 2012 17th International Conference on Computer Games (CGAMES). pp. 81–85. IEEE (2012)
11. Wiegers, K.: Software Requirements 2. Microsoft Press, Redmond (2003)
12. IEEE SA—830–1998—IEEE recommended practice for software requirements specifications, http://standards.ieee.org/findstds/standard/830-1998.html
13. Hunicke, R., LeBlanc, M., Zubek, R.: MDA: A formal approach to game design and game research. In: Proceedings of the AAAI Workshop on Challenges in Game AI. pp. 04–04 (2004)
14. Schell, J.: The Art of Game Design: a Book of Lenses [Paperback]. Morgan Kaufmann, Burlington (2008)

15. Boehm, B.W.: A spiral model of software development and enhancement. Computer (Long. Beach. Calif). **21**, 61–72 (1988)
16. Schwaber, K., Beedle, M.: Agile Software Development with Scrum. Pearson Education International, Boston (2002)
17. Beck, K.: Extreme Programming Explained: Embrace Change. Addison-Wesley Longman Publishing Co, Boston (1999)
18. Keith, C.: Agile Game Development with SCRUM. 1st edn. Addison Wesley, Boston (2010)
19. Kasurinen, J., Laine, R., Smolander, K.: How Applicable Is ISO/IEC 29110 in Game Software Development? In: 14th International Conference, PROFES 2013, pp. 5–19. Springer, Berlin (2013)
20. Godoy, A., Barbosa, E.: Game-Scrum: an approach to agile game development. IX SBGames. (2010)
21. Kortmann, R., Harteveld, C.: Agile game development: lessons learned from software engineering. In: Learn to Game, Game to Learn; the 40th Conference ISAGA 2009. Society of Simulation and Gaming of Singapore (2009)
22. Alexander, C.: The Timeless Way of Building: Oxford University Press, Oxford (1979)
23. Martín, D., Guzmán, J.G., Urbano, J., Llorens, J.: Patterns as objects to manage knowledge in software development organizations. Knowl. Manag. Res. Pract. **10**, 252–274 (2012)
24. Taylor, C.: Design template, http://www.runawaystudios.com/articles/chris_taylor_gdd.asp
25. Rogers, S.: Level up!: the guide to great video game design [Paperback]. Wiley, New York (2010)
26. Agile alliance: Continuos Integration, http://guide.agilealliance.org/guide/ci.html
27. Brinkkemper, S., Weerd, I., Weerd, S.: Developing a reference method for game production by method comparison. IFIP Adv. Inf. Commun. Technol. **244**, 313–327 (2007)
28. Wohlin, C., Runeson, P., Höst, M., Ohlsson, M.C., Regnell, B., Wesslén, A.: Experimentation in Software Engineering. Springer, Berlin (2012)

Emerging Industrial Visualization Systems: The Challenges and Opportunities of Virtual and Immersive Reality Technologies

Miguel Cota, Miguel González-Castro, Frederico Branco, Ramiro Gonçalves and José Martins

Abstract The continuous process industry is currently facing a serious need for systems that allow to monitor and manage their factories. Despite the existing industrial visualization screens provide for a partial answer to the identified issue, there is a set of new technologies who might help on developing the existing industrial visualization systems and provide and improved aid to the human controllers who monitor and manage the production processes. The presented paper aims at not only delivering a serious and focused characterization on the existing industry visualization screens, but above all, it intends to present several opportunities, challenges and new perspectives for the use of innovative visualization technologies (such as 2.5D/3D screens, augmented and virtual reality) in those industries whose production process assumes itself as continuous. This contribute, from our perspective will serve as a basis for further research on the field.

Keywords DCS · HCI · 3D · 2.5D · Smart glasses · Virtual reality, Augmented reality

M. Cota · M. González-Castro
University of Vigo, Vigo, Spain
e-mail: mpcota@uvigo.es

M. González-Castro
e-mail: miguelrgc@telefonica.net

F. Branco · R. Gonçalves · J. Martins
University of Trás-os-Montes E Alto Douro, Vila Real, Portugal
e-mail: fbranco@utad.pt

J. Martins
e-mail: jmartins@utad.pt

F. Branco · R. Gonçalves (✉) · J. Martins
INESC TEC and UTAD, University of Porto, Porto, Portugal
e-mail: ramiro@utad.pt

J. Mejia et al. (eds.), *Trends and Applications in Software Engineering*,
Advances in Intelligent Systems and Computing 405,
DOI 10.1007/978-3-319-26285-7_26

1 Introduction

Industrial visualization systems (IVS) have been evolving at the same pace that computers and information systems. The first IVS were introduced with the birth of Distributed Control System (DCS) [1], which are control devices that are mainly used in continuous process industry (e.g., steel, cement, paper). This type of industry is characterized by a critical production process, in which any failure or shutdown can create very hazardous situations.

The presented manuscript is the result of an effort towards identifying and highlighting the innovative perspectives and challenges associated with the use of virtual reality devices and immersive reality devices, in industrial environments where technicians, engineers and managers have a very critical need to access detailed information on the activities and processes that are being performed at the production plant.

In terms of presentation, the paper is divided in 5 sections, where the first is a straightforward description of the paper intent and structure. The Sect. 2 describes the methodology that supported the review to the IVS topic and that allowed to achieve a set of future challenges and perspectives associated with the emergent IVS. In Sect. 3 a presentation of DCS evolution supported by 2.5D/3D screens, smart glasses, and virtual and immersive reality technologies that help human controllers to perform their monitoring tasks in an easier and more efficient manner. The Sect. 4 encompasses the description of the set of future perspectives, challenges and uses for smart glasses, virtual reality systems and immersive reality systems, when applied to industrial organizations that are continuously producing. The Sect. 5 presents some conclusions and final considerations that summaries the presented manuscript.

2 IVS Characterization Approach

The level of knowledge associated with industrial management visualization systems is, at the light of our knowledge, not enough for IT managers and industrial production managers to draft and implement advanced solutions to not only monitor, but also to make real-time changes and configurations to industrial processes in order to maintain (or improve) the overall quality and efficiency. With this issue in mind, we decided to undergo an analysis to the existent industrial visualization systems and technologies in order to not only acquire the necessary know-how on the topic but to also develop our ability to suggest and/or implement something new to both the scientific and the industrial communities.

As a way to ensure scientific validity, and consequent contribute, of our efforts, we followed the approach indicated by authors such as Martins et al. [2], Oliveira and Martins [3], Morais et al. [4] and Kane et al. [5], who claim that in order to perform a structured characterization on a given topic one should mainly analyse scientific repositories, from which one should analyse the scientific and technical publications with the biggest impact and drilldown from there to other also important works. Drawing on those same authors, in this manuscript we present a very broaden description on all the reviewed works, with a more detailed attention to their technical and functional contributes.

3 Evolution of DCS Screens

Distributed Control Systems are platforms that manage thousands of analogue and digital signals which, through its distributed architecture, greatly increase its reliability and availability. Its main feature is the ability to deliver to their operators information on the industrial processes that it is controlling [6].

3.1 Current Industrial Display Systems

The optimal operator interface must provide an accurate and comprehensive "situational awareness" in all conditions status of industrial process (normal, abnormal and emergency) [7]. Therefore, the best DCS screen designs are those who optimize the quantity and quality of the industrial process information which the operator must assimilate and use to make the best decisions [8, 9].

The DCS operator interfaces that are sold today have 2D monitors and screens with process diagrams in 2D format [10]. This means that, if you want to have different views of the process, it is essential to create new screens. DCS alarms are indicated in the operation screen, with a change in the colour of the component that generates it, and also are listed in a table sorted by priorities or areas, which is accessible from different places of the screens. Also, the alarm occurrence causes the emission of an audible signal, composed of several tones.

Operators typically manage up to fifty DCS screens, each with several thousand I/O signals associated, thus allowing for the existence of displays or I/O signals that are not supervised by the operators during several work shifts, or until an alarm is activated. This highlights the need for summary displays where critical processes information are shown to the operator [11].

ABB and Umea Interactive Institute have developed a prototype of business management tool that facilitates the monitoring of any Key Process Indicator [12]. This management tool has a touch screen monitor which shows the production plant or industrial equipment in a 2.5D environment. The touch screen makes the rotation, scroll and zoom in the scene. The Siemens Comos-Walkinside application facilitates management tasks and engineering throughout the lifecycle of an industrial plant [13]. This tool maintains all the technical documentation of the plant.

3.2 DCS Operation Screens in 3D

Standard operation screens of large industrial automation systems have a flow diagram format in 2D. However, this format has severely restricted the amount of information it can hold. Likewise, this information should be well structured to facilitate the operator to understand it intuitively and with a single glance. This meant that new displays had to be developed in order to incorporate 2.5D and 3D features and allowing to present the information in a more understandable manner to the operator [6, 14].

Stereoscopic 3D visualization involves the operator to use both eyes for simultaneously perceive the 3D scene. Since the eyes are spaced a few centimetres between them, it causes that each eye perceives the scene from a different angle. The result is that the display of a real scene with both eyes adds a more visual perception of this scene. As a consequence, the information that an operator perceives in a 3D display is greater than if he was watching it in 2.5D.

Displaying 3D scenes requires the use of 3D devices in the operator station. These devices can be 3D glasses [15] or autostereoscopic displays [16–18]. As a requisite, 3D glasses should be of active type, and the operator must be able to perceive the colours and details of the graphic scene in extraordinary detail. The autostereoscopic displays require that the user is positioned exactly in the centre of the screen. However, this is not always possible, as the operator often change positions during his shift.

Cota and Castro [10] developed a software that can natively display operation screens in 3D, and if necessary, display them in 2.5D format on computers that do not have the necessary peripherals to display 3D images. Each of these new 2.5D and 3D operation screens may contain information that previously was shown on several operation screens. The application was tested on a DCS system that had several 2D screens and that monitored a set of paper past digesters from a paper paste industry. This test was with the collaboration of both the Grupo Empresarial ENCE (in Pontevedra, Spain) technicians and engineers and the researchers from the Department of Engineering of the University of Vigo. This company has installed a Honeywell DCS Alcont, composed by 2D screens, where each operator was responsible for operating multiple monitors, in which various industrial processes and operations were presented. Industrial process components could be selected, so that the available information is displayed in a small window command. This window provided a more detailed component information, including the operator full name and the operation/process state and the possibility to change it. Component selection was done with the mouse or by using an advanced pointer. The advanced screen pointer represents the geometric shape of octahedron and its movement is controlled via keyboard or 3D Joystick. This advanced pointer was the only available when viewing 3D scenes. By implementing this new 3D DCS, it was possible to perceive that the operator required less mental effort to understand the complete status of the industrial process. This allows to assert that the operation screen meets the key concepts of usability because it are very "user friendly" and very "easy to learn".

3.3 *Emerging Industrial Visualization Systems and Technologies*

By analysing the existing literature we were able to notice that some entities conducted studies on the use of advanced display systems in business, commercial and leisure activities. Muensterer et al. [19] presented an interesting contribute to the field of study, by describing the nine activities (GPS, video recording, transportation, construction, health, travel, education, security and advertising), where Google could have a major impact. The education sector is one area where more is innovating in the use of advanced display systems.

3.3.1 Smart Glasses

Smart glasses (Google Glass) are a display device that allows the user to view, on a single lens, an artificial image that is superimposed over his field of vision [20, 21]. Its development was focused on helping users to interact with their smartphones without using their hands and accept orders or verbal commands. This device incorporates several individual technologies, such as a gyroscope, a video camera, ambient light sensors, proximity sensors and bone induction system for the transmission of sound. It also allows communication via WiFi and Bluetooth.

Using Google Glass in industrial environments will allow to access information without using hands and simply by issuing verbal commands. This permits industrial workers to stay focused on the task they are doing with their hands while an enormous amount of information is superimposed on a small part of his field of vision. Bone induction system for sound generation also allow Google Glass to be used in association with sound protection systems, which are used in environments with high ambient noise. The use of these devices can be a good ally in preventing operator related accidents or errors, simply by reducing the unnecessary obligation to turn his head or move either hand to pay attention to information that is being received or to change the manual page you are reading. Finally, these display devices substantially reduce time spent gathering information prior and during the task being performed and the time needed to recover focus after task interruption.

Google Glasses are likely to be used for special applications, with a small adaptation in the peripherals you actually have installed. The installation of an infrared or a thermal camera facilitates the capture of a spectrum different from the scene which is captured. Infrared cameras can be of various types depending on if capture the near-infrared, medium or far away. Infrared spectrum image acquisition allows visualization of night scenes or scenes with much smoke or fog intense viewing. Thermal cameras are used to capture the thermal spectrum, to reveal the temperature of objects, or to differentiate objects based on its thermal structure. This represents an enormous aid in industrial environments, because at the same time that an operator sees a scene with different materials, he can also categorize it according to its internal structure (metal, wood, plastic, etc.).

Wille et al. [22] conducted a study that indicated that the tablet was more appropriate than Google Glass, as guide or handbook for the implementation of assembly tasks. However, these findings cannot be extrapolated directly to the tasks performed in industrial environments. This is because the study conditions and industrial environments are substantially different. Thus, Google Glasses allow the operator to have his hands free and to move without having to constantly holding or seeking the tablet. Finally, these glasses will not fade or constantly interfere with the operator's field of vision as it may be easily disconnect.

3.3.2 Augmented Reality

Augmented reality allows to add virtual information to the physical environment that is being observed. Technological devices facilitate augmented reality overlay of computer data on the actual scene the user is viewing. The magnified image is simultaneously composed of a real and artificial images. The artificial image is added to the original image in order to improve interpretation of the actual scene the user is receiving.

Smart glasses and augmented reality devices are based on similar technological principles. However, artificial images displayed in the lenses of both devices have different sizes, formats, and positions. Artificial images that are created in the augmented reality glasses are represented on the entire surface of the lens. On the other hand, the Google Glass project an image on a demarcated area of the surface of one of the lenses. Augmented reality devices are suitable for overlay information relative to the image that the user is receiving, without having to rotate the eyes and loose the focusing point. This allows information to flow naturally towards the user and don't need extra effort to understand what is being perceived.

Microsoft is developing a set of glasses called "HoloLens" [23], which combine augmented reality and 3D visualization. These goggles have transparent lenses on which holograms or 3D images are displayed and provide the user the feeling that the images are really embedded on the actual scene that they are viewing. The display of holograms provides the user with a sense of volume in the figures it receives and increases the conviction that these artificial figures are a part of the actual scene. The great advantage of using these glasses in any industrial environment is the fact that the user can see the holograms at the same time that it observes the actual environment in which he is in. Thus, the user is aware of the many dangers that exist around them. Similarly, in critical situations, an operator who carries these glasses and wear a protective suit, can reacquire all their visual field with a simple device that switch-off the glasses.

Augmented reality achieves data superimposed on a real scene to complement and help understand the displayed image, since the visual information of the scene which is perceived is improved. Thus, the user along with the visual image also perceives an interpretation of reality, which is due to the written information that appears on the screen. Similarly, if the information is not strictly complementary to the actual scene that is perceived, it should not be displayed or superimposed over the entire surface of the lens. For example, it would be inappropriate to superimpose a spreadsheet or pipes distribution chart on a real scene which is being observed. This type of information causes a huge confusion in the user, as text or graphics become unintelligible in some areas, as well as details of the actual scene are blurred. This information is best suited to be displayed in the box on the Google Glass. Much of the information displayed in the augmented reality devices can be adapted for display on the Google Glass; since in its projection box can show the real scene with a superimposed artificial image. However, very little of the information specially adapted for Google Glass can be viewed in augmented reality devices.

Augmented reality in industrial environments is a field that is still in the early days of its development. However, augmented reality devices will have a strong development in the very near future. Currently, both the scientific and business communities are starting to design the first prototypes [24], but it is our firm belief that these will shortly be used in a massive manner on industrial plants.

3.3.3 Immersive Virtual Reality

Virtual reality is a visual and sensory environment where scenes or artificial objects, which are generated by complex computing systems, acquire real appearance and the user has the feeling of being immersed in this virtual world [25]. This means that the user's senses are perceiving artificially created sensations, but that the user reads them as if they were real. This means that users will feel you are in a virtual world that is carried by its own senses.

Currently, are already being used virtual reality systems for training purposes. These simulators are also being used to train pilots and soldiers. Virtual reality transports the user to a world of illusions that, in some cases, may correspond to an existing reality, which can be relatively near to their physical location. This sensory illusion can be supplemented with information on the virtual scene being displayed. This will generate virtual scenes that are a hybrid between virtual reality and augmented reality.

4 Opportunities and Challenges

New advanced display systems will drive a major transformation to the industrial world in the way of aiding on the execution of maintenance, operation and design activities. Thus, industrial operators will have access to a universe of information that will allow them to optimize their work. Besides this, the operator can even transmit via video the task that he is performing and get help from an external team of experts, who are observing and analysing from a very remote place.

4.1 Industry Visualization Systems Current Concerns and Unsolved Issues

Next, we propose a set of perspectives and challenges susceptible of being associated with the different visualization equipment that was previously analysed, when applying them to industrial monitoring and management activities and processes.

4.1.1 Google Glasses

A very consensual use for Google Glasses is to use them to display operation and maintenance manuals and equipment schematics. These technical documents contain comprehensive information on all industrial processes that exist in the company and operational procedures. The visualization of the electrical or mechanical drawings at different scales and maps of the factory floor are critical, as they contain general information of all the factory and/or detail of each of the sections of the plant, and could also be incorporated and displayed to the user.

Operation screens can display the status of a facility or equipment. The lens box displays the contents of an operation screen, as this information enables the operator to make a safer work, informing him of hazardous locations, ensuring that orders and work permits maintenance to be performed the right way, looking at the checklist of tasks and connecting with ERP to perceive the availability of the material and make a request, if necessary, to the purchases department. Barcodes or RFID tags can be read for the identification of equipment, materials or products.

In noisy environments is possible to display alarm and signage, because information is always present in the field of view of the operator, prevents the need of the user to feel a vibration or hear an alarm signal. In an emergency situation the user/operator can also receive information about the location and routes to be followed to move towards emergency exits.

One of the most important tasks of any industry manager is reporting on meetings. This task is improved if the user has the most relevant information on the screen of his smart glasses. The operator can broadcast videos of work that is being carried out without any discomfort and without needing to carry any video recording device. This allows experts to give live indications or be recorded for later analysis.

4.1.2 Augmented Reality

Augmented reality equipment can be used to show hidden structures behind or inside a wall, equipment or bulky structures allowing, for example, its users to know which is the correct spot to drill a wall or present the values of the physical parameters of the equipment being configured (e.g., m3/h, amp).

Include information of the objects in the scene that is watching the user. Thus, to the scene you can add some arrows pointing to a device, along with a text box that contains the name of the component or any other important information for the user, or, additionally, identify the function of the keys or buttons on a keyboard, keypad, or any other user interface and warn the danger of a machine or installation by a colour code. Draw on the scene being observed, the route to be followed by the user to go from a source to a destination indicated by arrows on the scene, what is the way forward for the operator.

Overlay thermal infrared images on the actual scene the user is viewing allows, in smoky environments, to observe an infrared image of reality that is perceived by the user. This option is similar to one that is available for Google Glass but in these the image is displayed only in one quadrant. However, augmented reality enables infrared or thermal scene overlaps all the actual scene. This allows intuitively reveal a reality that was hidden from the operator.

4.1.3 Immersive Virtual Reality

Virtual reality systems are suitable for displaying buildings or industrial structures (e.g., beams, columns, tanks) while being designed, allowing early detection of problems that were not foreseen initially. They can allow de simulation of complex tasks performed by an operator to repair equipment (e.g., nuclear reactor, boiler and turbine), optimizing repair time and increasing their physical safety. These devices facilitate immersive knowledge of installations by the operator; allow testing different repair methods to optimize the task; and check the dangers of certain actions taken during this repair. This technology can be fundamental in the formation of technicians about processes and essential manoeuvres to be executed in critical or complex industrial equipment. Show manufacturing facilities to new operators, indicating how to reach the different areas of the factory. Also the location of different equipment or machines in the factory and how to get to those places.

Previewing new industrial products right from the design phase will certainly improve the product overall quality and adjustment to the customer needs, given that it would allow for a better analysis of the technical characteristics which they must comply. This virtual decomposition of a complex product in simpler pieces and the ability to adjust each peace to improve its viability is one of the most important features associated with immersive virtual reality systems. Moreover, these virtual design, customization and assembly stages will improve the product real installation but also its repair operations.

Overlapping real data of the industrial process to real images of machines or instruments, allows process operators to know the state of the plant and even get to operate it. This operation is suitable for industrial equipment that is in the virtual surrounding place where the operator is located. If real equipment that had to be handled are much distanced from the other, the operator should traverse a virtual path to go from one to the other industrial device. These tours of the operator, even if virtual, are critical as, while it invests in this virtual displacement, result in real time. This would be unacceptable in critical situations, in which the speed of the operator to activate/deactivate the appropriate equipment is critical to resolve this troublesome condition. Also, the display of the actual images of the process make it difficult for the operator to have a full understanding of the situation of the whole industrial process that is managing. Finally, a display of real images of the industrial process prevents the creation of screens summary, containing the main data of the process.

5 Conclusions and Final Considerations

When reviewing the existing literature that focused its attention on IVS systems one can easily identify a knowledge gap that we aimed to address with the present research. As the result of the performed characterization of IVS topic, four main technologies have arisen as being the ones with the most potential to incorporate in industrial visualization systems: (1) 3D DCS operating screens, (2) smart glasses, (3) augmented reality, and (4) virtual reality. To our knowledge, as the result of the combination between their characteristics and their functional features, these are the technologies with the most innovative and success perspective in what concerns the development of IVSs.

The industrial visualization systems are a key part in monitoring and management activities of industrial organization, mainly because they allow for a proper comprehension of the productive processes and consequently to make accurate and informed decisions. Through the study of the four presented technologies, the research team acknowledged that there is a long path to be persecuted in order to associate the technologies benefits and the monitoring and management needs inherent to modern industries.

References

1. Pauly, T.: Distrib. Control System Electron. Power **33**, 573–576 (1987)
2. Martins, J., Gonçalves, R., Oliveira, T., Pereira, J., Cota, M.: Social networks sites adoption at firm level: a literature review. In: CISTI'2014—Iberian Conference on Information Systems and Technologies, Barcelona, Espanha (2014)
3. Oliveira, T., Martins, M.: Literature review of information technology adoption models at firm level. Electron. J. Inf. Syst. Eval. **14**, 110–121 (2011)
4. Morais, E., Gonçalves, R., Pires, J.: Electronic commerce maturity: a review of the principal models. In: IADIS International Conference on e-Society, Lisbon, Portugal (2007)
5. Kane, G., Alavi, M., Labianca, G., Borgatti, S.: What's different about social media networks? A framework and research agenda. MIS Q. (2012)
6. Gonçalves, R., Martins, J., Branco, F., Castro, M., Cota, M., Barroso, J.: A new concept of 3D DCS interface application for industrial production console operators. Universal Access in the Information Society, pp. 1–15. (2014)

7. Bamieh, B., Paganini, F., Dahleh, M.: Distributed control of spatially invariant systems. IEEE Trans. Autom. Control **47**, 1091–1107 (2002)
8. Reising, D., Laberge, J., Bullemer, P.: Supporting operator sitation awareness with overview displays: a series of studies on information vs. vitualization requeriments. ICOCO, pp. 188–198 (2010)
9. Breibvold, H., Olausson, M., Timsjo, S., Larsson, M., Noble, R., Sneddon, N.: El operario eficaz. Revista ABB **4**, 6–11 (2010)
10. Cota, M., Castro, M.: DCS 3D operators in industrial environments: new HCI paradigm for the industry. Virtual, augmented and mixed reality. systems and applications, pp. 271–280. Springer (2013)
11. Sharma, C., Bhavsar, P., Srinivasan, B., Srinivasan, R.: Understanding cognitive behavior of process operators during abnormal situations through eye tracker studies. AIChE Annual Meeting. AICHE, Atlanta, USA (2014)
12. Olausson, M., Larsson, M., Alfredsson, F.: Colaborando en una nueva dimensión: Una presentación interactiva ayuda a tomar las decisiones correctas. Revista ABB 6–11 (2012)
13. Payne, A., Frow, P.: A strategic framework for Customer Relationship Management. J. Mark. **69**, 167–176 (2005)
14. Cota, M.P., González-Castro, M.R.: Usability in a new DCS interface. Universal Access in Human-Computer Interaction. Design Methods, Tools, and Interaction Techniques for eInclusion, pp. 87–96. Springer (2013)
15. http://www.nvidia.es/object/3d-vision-pro-requirements-es.html
16. Dodgson, N.A.: Autostereoscopic 3D displays. Computer, vol. 38, pp. 31–36. (2005)
17. Urey, H., Chellappan, K.V., Erden, E., Surman, P.: State of the art in stereoscopic and autostereoscopic displays. Proc. IEEE **99**, 540–555 (2011)
18. Karaman, E., Cetin, Y., Yardimci, Y.: Angle perception on autostereoscopic displays. In: 3rd Conference on Human System Interactions (HSI), pp. 300–307. IEEE (2010)
19. Muensterer, O., Lacher, M., Zoeller, C., Bronstein, M., Kübler, J.: Google Glass in pediatric surgery: An exploratory study. Int. J. Surg. **12**, 281–289 (2014)
20. Triggs, R.: How it works: Google glass. Android authority, http://www.androidauthority.com (2013)
21. Google: Google glass (2015)
22. Wille, M., Scholl, P., Wischniewski, S., Van Laerhoven, K.: Comparing google glass with tablet-pc as guidance system for assembling tasks. In: 11th International Conference on Wearable and Implantable Body Sensor Networks Workshops (BSN Workshops), pp. 38–41. IEEE, Zurich, Switzerland (2014)
23. http://www.microsoft.com/microsoft-hololens/en-us
24. Perrigot, R., Kacker, M., Basset, G., Cliquet, G.: Antecedents of early adoption and use of social media networks for stakeholder communications evidence from franchising. J. Small Bus. Manage. **50**, 539–565 (2012)
25. Seth, A., Vance, J., Oliver, J.: Virtual reality for assembly methods prototyping: a review. Virtual Real. **15**, 5–20 (2011)

Author Index

J. Mejia et al. (eds.), *Trends and Applications in Software Engineering*,
Advances in Intelligent Systems and Computing 405,
DOI 10.1007/978-3-319-26285-7

Printed in the United States
By Bookmasters